Defining Species

AMERICAN UNIVERSITY STUDIES

SERIES V
PHILOSOPHY

VOL. 23

PETER LANG
New York • Washington, D.C./Baltimore • Bern
Frankfurt am Main • Berlin • Brussels • Vienna • Oxford

John S. Wilkins

Defining Species

A Sourcebook from Antiquity to Today

PETER LANG
New York • Washington, D.C./Baltimore • Bern
Frankfurt am Main • Berlin • Brussels • Vienna • Oxford

Library of Congress Cataloging-in-Publication Data

Wilkins, John S.
Defining species: a sourcebook from antiquity to today / John S. Wilkins.
p. cm. — (American university studies. V, Philosophy; v. 203)
Includes bibliographical references and index.
1. Species—Philosophy. I. Title.
QH83.W526   578.01'2—dc22   2008027030
ISBN 978-1-4331-0216-5
ISSN 0739-6392

Bibliographic information published by **Die Deutsche Bibliothek**.
Die Deutsche Bibliothek lists this publication in the "Deutsche
Nationalbibliografie"; detailed bibliographic data is available
on the Internet at http://dnb.ddb.de/.

The paper in this book meets the guidelines for permanence and durability
of the Committee on Production Guidelines for Book Longevity
of the Council of Library Resources.

© 2009 Peter Lang Publishing, Inc., New York
29 Broadway, 18th floor, New York, NY 10006
www.peterlang.com

Printed in Germany

To David Hull, who has always helped
To Paul Griffiths, who gave me a chance
And to my family, who have always understood

Contents

❧ Acknowledgments

This book is not my own work alone. I am deeply indebted to the following for criticism, editing, material, suggestions and the occasional expression of outright incredulity:

In alphabetical order, Floyd Aranyosi, Chris Brochu, Michael Dunford, Malte Ebach, Greg Edgcombe, Dan Faith, Michael Ghiselin, Paul Griffiths, Colin Groves, John Harshman, Jody Hey, David Hull, Mike S.-Y. Lee, Murray Littlejohn, Brent Mishler, Larry Moran, Staffan Müller-Wille, Ian Musgrave, Gary Nelson, Mike Norén, Gordon McOuat, Massimo Pigliucci, Tom Scharle, Kim Sterelny, Neil Thomason, John Veron, David Williams, and Polly Winsor, who all provided information, advice and assistance, some considerable;

Polly Winsor, David Hull, Joel Cracraft, David Williams, Norman Platnick and Ward Wheeler also graciously granted time for an interview. I must also thank the late Herb Wagner, John Veron, Mike Dunford, Tom Scharle, and Scott Chase for technical information provided. Other acknowledgements are made in the notes. I sincerely apologise for anyone I have left unthanked.

Therefore, in the light of this generous assistance from so many people, it follows that all misunderstandings, errors and incoherencies that remain are my own fault.

Original texts are often available from the *Gallica* project at the Bibliothèque nationale de France <gallica.bnf.fr>, the *Internet Archive* <www.archive.org>, and Botanicus <www.botanicus.org/>, to which projects I offer my gratitude. Also, the *Complete Works of Charles Darwin* project <darwin-online.org.uk>, overseen by John van Whye, is an invaluable resource. All quoted texts are the copyright property of their owners, and are used under "fair use" except for the permissions given below.

The following material is reprinted with permission. All attempts have been made to contact copyright holders, and gain permission for material not covered by fair use. If you believe material for which you hold copyright is reprinted here and is not covered by fair use provisions, please contact the author, John Wilkins, at john@wilkins.id.au and arrangements to pay permission costs will be made where appropriate.

Other permissions are noted in the text.

৯ Introduction

Of all histories the history of ideas is the most difficult and elusive. Unlike things, ideas cannot be handled, weighed, and measured. They exert a powerful force in human history, but a force difficult to estimate. [(Greene 1963: 11)]

Definitions of species (Greek, *eidos* [εἶδος], Latin, *species*) go back to the Greeks, and in particular Plato, whose definition of Form (which also translates *eidos*) included it being unchanging, separated from all other forms, and having the same essence in all cases of it. But Plato's form was not a material thing, and it applied equally to society as it does to the natural world. Aristotle's subsequent understanding of form, though, was of something that existed solely in material things. Aristotle's use of *eidos* was a logical use—each kind of thing (*genos* [γένος], *genus* in Latin) could be divided into further kinds (*eidē*) that shared the general properties of the genus, but which had specific differences (*diaphora* [διαφορά], *differentia* in Latin) that separated them. Aristotle's logic is the *logic of division* or, in Greek, *diairesis* (διαίρεσις). Henceforth, we shall refer to it as the *diairesis*.

Aristotle did not apply this within natural history solely to kinds of living organisms, but instead also to kinds of organs, and organisms of a very broad variety depending on what was being discussed (for instance, habitat or structure). The medieval development of *logical species* was broadly based on Aristotle's views, whether via Porphyry and Boëthius in the early part of the period, or via Michael Scot's and subsequent translations in the latter period.

The consistent use of *species* to refer to the lowest kind of living organisms, which would have been *infimae species* (bottom kinds, or least inclusive kinds) in logic, did not develop until the 17th century, and in particular with John Ray's definition. Thereafter, *species* in natural history took on a life of its own, as the term had previously done in psychology and theology. It is not appropriate to take logical definitions as automatically applying to *natural species* (also sometimes called "organised beings") unless the author directly does this. To the contrary, many natural historians and logicians overtly noted the distinction between logical and natural species. It is also worth noting that the terms *genus* and *species* were also "vernacular" terms in Latin throughout this period, and were often used in a non-technical sense even when the authors knew of the logical and natural definitions, for stylistic considerations, such as avoiding repetition of words.

I have divided the early material into three periods: pre-modern, early modern and modern, broadly being divided by focus on the traditions from the Greeks, the rise of natural history, and biology after Darwin. Darwin is significant not so much because he represents any decisive break in biology about species, although in many other senses he does, but because biologists and historians treat him as a turning point, and the present debate refers to him. After the introduction of the so-called modern species debate by Dobzhansky, I divide the conceptions into the major categories into which they fall—reproductive isolation, evolutionary, phylogenetic, ecological, asexual, and "other". In many cases, a particular definition properly covers more than one of these categories, but I give them in each case in their primary group. The final section covers commentary and philosophical views on species definitions.

The history of the *species* concept can be divided into a pre-biological and a post-biological history, which is how the received view has always treated it. But the two histories overlap substantially, and it is much better to consider instead the history of the species idea that applies to any objects of classification—the tradition of universal taxonomy and philosophical logic—and, independently, the particular history of the species idea that applies solely to biological organisms. Even though, for example, Linnaeus (Linne 1788–1793 vol. 3) famously applied the notion *species* to minerals as well as organisms, his biological use of the term included elements not included in the mineralogical case. We must separate universal and biological taxonomic notions of *species*.

Therefore, we must distinguish between two kinds of taxonomy. The *universal taxonomy* is largely the philosophical tradition from Plato to Locke (but which continues through to the considerations of sensory impressions, or *qualia*, by the logical positivist and phenomenological philosophical schools) and in which species are any distinguishable or naturally distinguished categories with an essence or definition. Then there is the *biological taxonomy* that develops from this tradition as biology itself develops from the broader field known as "natural history".[1] These biological notions of species do not necessarily refer to reproductive communities, nor do they in the medical definitions of species of

---

1   "History" is, in this older sense, the Greek word *historia*, which means an inquiry or investigation, which derives from the title, or rather the opening words, of Herodotus' *History*. Later it comes to mean "knowledge" or "learning".

diseases of the period (Cain 1999), but we do need to recognize that "species" develops a uniquely biological flavour around the seventeenth century.[2]

These two conceptions form what might be regarded as part of or an entire research program, in Lakatos' sense. The universal taxonomy program that began with Plato culminates in the seventeenth century with the attempt to develop a classification of not only all natural objects, but all possible objects. It continues in some sense in the modern projects of metaphysics and modal logics. The biological taxonomy project that developed out of it resulted in a program to understand the units of biology, to which Darwin contributed, and to which genetics later added. We might see *species* as a marker for both projects and thus reasonably ask what the "core assumptions" of these projects are. In the universal case, it is classification (and science) by division (of terms). In the biological instance, it is the marriage of reproduction or generation, with form. This remains even today the basis of understanding *species*. It is important to realise that the use of logical terms does not commit naturalists and biologists to the definitions of logic in natural history or taxonomy. In fact, nearly all naturalists understood that logical definitions were not transferable to biological species. And most of them realised that what made a species was the propagation of like kinds through some sort of generative power, usually involving interbreeding, which I call the *Generative Conception of Species*.

My target here is the so-called "Essentialism Story" (Amundson 2005; Levit and Meister 2006; Winsor 2006a, 2006b), largely devised by Ernst Mayr, in which species before Darwin are supposed to have been defined by a fixed and untransmutable essence. I believe this is historically wrong. Few if any held that essences prohibit transmutation of species but rather that fixity was required by doctrinal orthodoxy and piety, not metaphysics. Moreover, fixity was not held by many naturalists before Ray, if any. Since Ray effectively established a natural species concept, however, fixism may be taken to be the pre-evolutionary (not pre-Darwinian as such) conception of 'species" in the period from around 1735–1860.

Terms and Concepts

The modern and medieval word "species" is a Latin translation of the classical Greek word *eidos*, sometimes translated as "idea" or "form". Other significant

---

2   An excellent treatment of many of these themes, and a good aid to understanding the historical contexts, can be found in Mary Slaughter's wonderful book (1982); a broader and more liberal treatment, but one that suffers from over-theorizing, is Michel Foucault's *The Order of Things (Les Mots et les Choses)* (Foucault 1970), particularly chapter 5.

terms are also translations of Greek terms: *genus* from *genos*, *differentia* from *diaphora*. Liddell and Scott (1888) tell us that *eidos* means "form", and is derived from the root word "to see", and *genos* means "kind" and is derived from the root word "to be born of". We still find these senses in the English words "specify", "special", "spectacle", and "generation", "gene" and "genesis". *Eidos* was originally associated with visible form or appearance. *Genos* was originally a tribe or family, which therefore resembled each other. We shall see that these were often used interchangeably in ordinary discourse, despite being "technical terms" in post-Aristotelian logic. In fact, Aristotle himself used them in the vernacular sense in his biological texts.

However, merely because words derive their etymology from older terms, or translate words in other languages, it does not immediately follow that they are the same terms with the same intension or extension, or that one author is influenced by another. By *eidos* and *species*, for example, different authors have meant *forms*, *kinds*, *sorts*, *species* (in the logical sense of non-predicables, Ross 1949: 57), *biological species*, *classes*, *individuals*, and *collections* (arbitrary and artificial, or natural and objective). In particular, *species* has meant varieties of ideas and sense impressions, *species intelligibilis*, (Spruit 1994–1995) and also the material form of the elements of the sacraments in the Roman Catholic tradition. More recently, the mineralogical use has been retained in chemistry, where "chemical species" are often discussed.

This must be borne in mind, or it will cause confusions when considering the views of different authors. This is an instance of a more general problem, called "incommensurability" after Kuhn's thesis that terms in scientific theories can have different referents in the shift from one theory to another; Sankey (1998) has called this particular problem "taxonomic incommensurability". We should not make too much of this, but the terms used in classification shift in subtle and major ways that sometimes obscure the views each author is presenting. We are primarily concerned with the tradition outside biology that *has* impacted on the biological notions and usage.

The English plural of "species" singular is "species". The word "specie" refers to small coinage. However, in Latin there is a singular (*species*) and plural form (*speciei*), which is signified in one text (Porphyry 1975) by italicising the ending: spec*ies*. The rule of the biological writers of the past century is to refer to "the" species or italicise the entire word for the concept, and leave the term unqualified for number except by context.

For the period from the Greeks to the beginnings of natural history in the 16th and 17th centuries, it may help to follow Locke's suggestion, to get a feel for the meanings and avoid anachronistic interpretations, by replacing *genos* and *genus* with "kind" in English, and *eidos* and *species* with "sort". Thus, the informal usage of, say, Theophrastus in talking indifferently about *genē* and *eidē*, can be seen as the same informal usage an English speaker might use by stylistically mixing "kind" and "sort" in a discussion to avoid repetition. It is imperative to remember that these were not technical terms of biology until the modern period, in particular after Linnaeus. Likewise the terms *genus* and *species* were, as well as being technical terms of art in logic, ordinary parts of the vocabulary of Latin speakers, right up until the modern period. Not every use of these terms indicates a metaphysical or logical commitment.

It is important also to recognise that there is one *concept* of (living) species—the concept that living things are organised into a smallest formal group or class. What are typically referred to as "species concepts" are in fact definitions or explications of *the* species concept according to the author quoted. To overcome this confusion, Mayr distinguished between species as taxa and species as a category, but this is only because he treated his own definition as a separate concept. So here, I call them "conceptions", or definitions, of the concept, following the older, preferable, pre-Mayrian, practice, and in line with James Mallet's point (Mallet 1995).

A Note on the Choice of Selections

The literature on definitions of species is almost fractally complex. For almost every text, there are other texts that clarify, dispute or elaborate it. A complete set of texts would run to many volumes of largely repetitive and often dull prose. My choice of selection is therefore motivated by my conclusions about the history of the species concept, and by the fact that certain definitions and discussions are more influential in terms of being cited, than others. It is, in other words, a start, not an end.

Many texts are widely quoted because they are the texts that later influential authors first encountered. Some texts which have been influential are hardly ever quoted, though they changed the way the debate was framed at the time (for instance, Poulton or Gilmour). Others I include because they show that the way in which the Essentialist Story interprets them is misconceived.

For the post-Synthesis period, because so much has been published on species conceptions, and because the material is fairly readily available, I have only given the initial definitions and discussions rather than try to collect all the ones in the present literature, which would require an encyclopedia rather than a reader.

If you think I have left out some important passage (not from textbooks unless that formulation is influential), or misinterpreted a passage or author, please contact me at john@wilkins.id.au and let me know the source and text. I can guarantee on a priori grounds that there will be many cases of this.

Typographical Note

All em dashes in the original texts have been converted to an en dash without spaces. Spellings are left as is. All other amendments have been noted in the text. Notes by myself have been marked in square brackets and comments identified with "*JSW*". Original pagination breaks in long texts is shown in square brackets also. Footnotes in the original text have been included in the quoted material after the paragraph or page in which they occur.

❧ Section 1. The Pre-Modern Definitions

In the classical period from Plato to Boethius, "species" typically means something like a subset of a larger set, in modern terms. The *genus* is the "general predicate", that is, the broader class, which is divided into the *species*, or "special" predicate. This is applied indifferently to social, moral, aesthetic and natural kinds. The logic of division, or *diairesis*, which Aristotle developed formally, remained largely unchanged until the 19th century, although it is clear that in the late classical period, and to an extent in the early middle ages, this logic was misunderstood and often confused by writers. Not until the translation of many of Aristotle's works in the 11th century did logic become well studied again, finding its way into the ordinary educational curriculum of scholars known as the Trivium.

Later, Catholic doctrine, especially on the Incarnation and the Sacrament of Communion, caused confusion of Aristotelian logic (and metaphysics) by trying to use concepts in a way that was at odds with the *diaresis*. So it is not unusual to find confusions in a purely technical sense amongst theologians as late as the turn of the 20th century.

No separate program of classification of living things from other things was begun until the early modern period. However, as evidenced by Frederick II below, practical experience often led to an understanding of species as generative. Throughout the medieval period, however, lists of beasts (in bestiaries) and of herbs (in herbals) had other purposes than taxonomy. Bestiaries were often fabulous and moral homilies, while herbals were local summaries of pharmacological and gastronomic uses. The use of species could be a variety, in modern terms, or a genus or higher. However, as Stannard has ably shown (Stannard 1968, 1979, 1980b, 1980a, 1999; Stannard, Kay, and Stannard 1999), such collations often closely tracked species, at least locally.

Nonetheless, the so-called Aristotelian program was based on the idea that the knowledge embedded in ordinary language could, with a bit of logical analysis, render scientific knowledge. In other words, Aristotle thought to some extent that one could do science by definition (e.g., *Topica* Bk I, ch 5, 101^{b}37ff), despite his empirical researches and use of observations by travellers.

Plato (c.428/427BCE–c.348/347BCE)

Plato did not write much about living kinds—he was primarily interested in human society and moral questions when it came to classification. But the comment from the *Phaedrus* below has been widely quoted in taxonomic contexts, although it refers to what we would think of as a social, not a natural, kind. Classification was at this time uncritically applied to all things, whether artificial or natural (a distinction the early Greeks would not have fully accepted anyway) and whether conceptual, semantic, or empirical. The problem of how to properly classify things, including living things, is first recorded to be dealt with in detail by Plato in the *Sophist*. Plato, regarded by many later thinkers as *The* Philosopher, founded the philosophical school known as the Academy in Athens. He proposed a method of binary division of contraries until the object being classified was reached; this was known as the *diairesis* (division, or as some call it, dichotomy), in a similar fashion to the Pythagoreans.[3] For example, he somewhat whimsically defined fly-fishing as a model for all such classification. He has the Stranger of the dialogue ask leading questions, such as whether the fisher has skill (*techne* = art) or not, defining art into two kinds, agriculture and tending of mortal creatures on the one hand, and art of imitation on the other, and later introduces a fairly arbitrary distinction about acquisitive art, resulting in the following "final" definition of angling:

> STRANGER: Then now you and I have come to an understanding not only about the name of the angler's art, but about the definition of the thing itself. One half of all art was acquisitive—half of the acquisitive art was conquest or taking by force, half of this was hunting, and half of hunting was hunting animals, half of this was hunting water animals—of this again, the under half was fishing, half of fishing was striking; a part of striking was fishing with a barb, and one half of this again, being the kind which strikes with a hook and draws the fish from below upwards, is the art which we have been seeking, and which from the nature of the operation is denoted angling or drawing up (*aspalieutike, anaspasthai*). [*Sophist* 219a–221a: Jowett's translation]

---

3   Some believe that Plato was a Pythagorean who broke the mystery boundaries of the religion, and at least one (John Bigelow, pers. comm.) argues that some of Plato's work, like the *Phaedo*, is a guarded expression of Pythagorean mysteries regarding the ratio, or *logos* of number.

Soc. The second principle is that of division into species according to the natural formation, where the joint is, not breaking any part as a bad carver might. [*Phaedrus* 265d–266a: Jowett's translation]

Aristotle of Stagira (384BCE–322BCE)

Aristotle had several projects, one in logic and metaphysics (which are discussed in the *Posterior Analytics* and the *Metaphysics*), as well as a series of books about zoology, as he conceived it—the trilogy called the *Liber Animalium*, which comprises *History of Animals*, *Generation of Animals* and *Parts of Animals*. Other works that are relevant to his biology, but not particularly on this topic here, include *On the Soul*, *On Generation and Corruption*, *On the Movement of Animals*, and *On the Progression of Animals*.

It is hard to summarise Aristotle's views simply. His logical project proceeded by dividing already known general terms ("predicates") into increasingly specific analytic components until you had specified all that you could about the object and only now count different instances. So it appears that he thought science could be done, indeed should be done, by definition or division.

And yet, he undertook empirical research on his own account, and with students and travellers, on the diversity of life in form and function and distribution. So it appears that he did not dogmatically require that every bit of knowledge must be got from ordinary language.

Aristotle's main target here is the classification by dichotomy into predicates and their opposites that Plato and the members of the Academy adopted. He thinks that general predicates, or *gene*, can be divided into many subgroups, and that division into what came to be called *privative*, or negative, groups caused incoherence in the classification. Hence, a *genos* is divisible into as many *eide* as there are differentiating characters. Also, though, he needs to identify the proper characters, and not the incidental ones, and so he distinguishes between the "what it is to be" characters, in Latin, essential characters, from the "accidental" (*symbebekos*) ones that can be ignored or changed without affecting the classification. In the *Topics* (101b16–25) Aristotle defines four "predicables" (that which is predicated of things): definition (*horos*), property (*idion*), genus and accident (*symbebekos*). Species (*eidos*) is not, in Aristotle's list, a predicable, because it is only true of individuals.

He extends his discussion in the *Metaphysics* in book Z, chapter 12 (1037b–1038a), by asking what it is that makes *man* (the logical species) a unity instead of a "plurality" such as *animal* and *two-footed*. He argues that the differentiae of a genus can lead to its including species which are polar opposites in their specific differences, but, with respect to the genus itself, there is no differentiation. This makes sense only if each genus is divided further in terms other than the predicates that define the genus. Further, he rejects Plato's dichotomous approach, saying "… it makes in general no difference whether the specification is by many or few differentia, neither does it whether that specification is by a few or by just two… ." Therefore, he asks whether the genus exists "over and above the specific forms constitutive of it", and answers that it doesn't matter, because the definition is "the account derived just from the differentiae". In the end, we reach the "form and the substance", the last differentia, unless we use accidental features, in which case we will find that we have an incorrect division as evidenced by the differentiae being "equinumerous with the cuts". In short, a species is the form and the substance of the genera, when we reach the last differentia.

It is sometimes held that for Aristotle all classification was in terms of absolute definition and essence, and often this is true. But he did allow for an excess or deficiency of some organs or properties of organisms in the *Parts of Animals* (e.g., 646b 20) and in the *Rhetoric*, one of the main topics or lines of argument was "the more and the less" (1358a21). In the eighth book of the *Metaphysics* (Book Iota), he even says

"… it is important to understand the kind of differentiation, given that they are the principles of the beings of things. Some things, that is, are marked out by being more or, conversely, less F, by being dense, say, or rare and so forth, which are all instances of the surfeit/deficiency differentiation." [1042b, Lawson-Tancred translation.[4] (Aristotle 1998)]

Similarly, in the *History of Animals* (486a–486b) he discusses differences being more or less the same property in respect of the genus, and "in short, in the way of excess or defect", "for the more and the less may be represented as excess and defect". (D'Arcy Thompson translation, Barnes 1984). Species of birds and fish, for instance, may not properly instantiate their genus. The more and the less, however, refer to aspects of the *eidos* that can vary over a range, and which can be important for the organism's life (Lennox 2001: 178). The range is precise, and forms part of the differentia of that species. In the *The Parts of Animals*, Aristotle discusses why privative terms are not proper to classification (book I, chapter 3, 642b–643a). He says that you cannot properly further divide a privative term, and that precludes it from being a generic (that is, general) term. Worse, a privative classification can include the same group

---

4   In other translations, such as W. D. Ross's (in McKeon 1941), the phrase is " the kinds of differentiae .. characterized by the more and the less".

under contradictory terms. Classifications, according to Aristotle, must say something direct and clear. Dividing the world into things that are, and aren't, describable by some predicate, is at best only a partial classification, and the taxa that result are not good divisions of the broader genus. Some things that fall under a privative term must be species, but there is no genus out of which those species can be differentiated, and so the privative "genus" is illusory. Plato's mistake was, he thought, to assume that classifications must be made in terms of polar opposites.[5] Aristotle was saying, as we would now describe it, that classification must always be cast in terms of proper sets and subsets. Partial inclusion is not legitimate in a good classification, in effect because it does not make complete sets.

It is traditionally held that Aristotle was inconsistent in the way he used *genos* and *eidos* between the logical and the biological writings (e.g., Mandelbaum 1957) but recent work by Pellegrin, Balme and Lennox has shown otherwise (Balme 1987a; Lennox 1987; Lennox 1993, 1994, 2001; Pellegrin 1986, 1987). In part, the problem arises because the common view rests mainly on the post-medieval concepts that are derived out of the later neo-Platonic revision of Aristotelian logic. Aristotle is only inconsistent if understood to use the term *eidos* in the same way that Porphyry and others later use the term *species*.

Pellegrin says that Aristotle was not aiming to produce a biological taxonomy in *History of Animals*. Instead, he was producing general classifications, and animals happened to be one domain in which he applied that method. What Aristotle treats as genera and species do not answer directly to the modern, post-Linnaean, conceptions of species, although this has sometimes been the default interpretation. We have seen that for him a species is a group that is formed by differentiating a prior group formed by a generic concept. Genera have essential predicates (or definitions), and so do species. Infimae species happen to be indivisible, that's all. In this respect, biological species are no different to any other kind. Pellegrin (p110) says, "Aristotle thus conveys by the term *genos* the transmissible type that in our eyes characterises the species, and by *eidos* the model that is actually transmitted in generation. It would be necessary for these two terms to converge and become superimposed for the

---

5   Balme says that in *Parts of Animals* 1.2-4 (Balme 1987b: 19):
     … Aristotle concludes that diairesis can grasp the form if it is not used dichotomous-ly as Plato used it but by applying all the relevant differentiae to the genus simultane-ously; after that he explains the ways in which animal features should be compared so as to set up differentiae—by analogy between kinds (genē), by the more-or-less as between forms comprised within a kind (eidē).

modern concept of a species to be born. For Aristotle, the species did not yet exist."

Still, it is tempting to see Aristotle as primarily an ordinary language philosopher of the kind exemplified by Ryle, Austin, and Strawson. For instance, his notion of *katholou*, later translated as "universal", is an ordinary word of Greek pressed into philosophical service: it is a portmanteau word formed from the Greek *kata*, meaning "according to" or "in terms of", and *holou*, meaning "of the whole". A universal is a term that is predicated "according to the whole". Similarly, the terms he used in his logical and metaphysical work were not neologisms. They had prior usage, going back at least to Homer in many cases, and had connotations of a particular kind that they lack in their Latin equivalents.

So, it's necessary to set up a translation table between the original Aristotle, if he were available to us, and the Latinised Aristotle of the scholastics. For that reason I use the G. R. G. Mure translation, with its use of medieval terms, for the first excerpt, and the translations of the Barnes edition in subsequent ones.

Infimae species and division of topics

The authors of a hand-book on a subject that is a generic whole should divide the genus into its first *infimae species*—number e.g,. into triad and dyad—and then endeavour to seize their definitions by the method we have described ... . After that, having established what the category is to which the subaltern genus belongs—quantity or quality, for instance—he should examine the properties 'peculiar' to the species, working through the proximate common differentiae. He should proceed thus because the attributes of the genera compounded of the *infimae species* will be clearly given by the definitions of the species; since the basic element of them all [note: *sc.* genera and species] is the definition, i.e. the simple *infimae species*, and the attributes inhere essentially in the simple *infimae species*, in genera only in virtue of these. [*Posterior Analytics* 96^{b}15–24 (McKeon 1941)]

... a thing is not different in virtue of every difference; for many differences belong to things that are the same species—though not in respect of their substance, nor in themselves.

Next, when you assume the opposites and the differentia and that everything falls here or there, and assume that what you are seeking is in one of them, it makes no difference whether you know or do not know the other things of which the differentiae are predicated. For it is evident that if, proceeding in this way, you come to things of which there is no longer a differentia, you will have the account of its substance. ... [*Posterior Analytics*, 97ª12–20 (Aristotle 1995)]

To establish a definition through divisions, one must aim for three things—grasping what is predicated in what the thing is, ordering these as first or second, and ensuring that these are all there are. [*Ibid*, 97ª24–26]

Every definition is universal; for the doctor does not say what is healthy in the case of some individual eye, but either in the case of every eye, or determining some species of eye. [*Ibid*, 97ᵇ25–27]

Substance and essence

Since the subject of our inquiry is substance, let us return to it. Just as the substrate and the essence and the combination of these are called substance, so too is the universal. With two of these we have already dealt, i.e. with the essence and the substrate; of the latter we have said that it underlies in two senses—either being an individual thing (as the animal underlies its attributes), or as matter underlies the actuality. The universal also is thought by some to be in the truest sense a cause and a principle. Let us therefore proceed to discuss this question too; for it seems impossible that any universal term can be substance.

First, the substance of an individual is the substance which is peculiar to it and belongs to nothing else; whereas the universal is common; for by universal we mean that which by nature appertains to several things. Of what particular, then, will the universal be the substance? Either of all or of none. But it cannot be the substance of all; while, if it is to be the substance of one, the rest also will be that one; because things whose substance is one have also one essence and are themselves one.

Again, substance means that which is not predicated of a subject, whereas the universal is always predicated of some subject.

But perhaps although the universal cannot be substance in the sense that essence is, it can be present in the essence, as "animal" can be present in "man" and "horse." Then clearly there is in some sense a formula of the universal. It makes no difference even if there is not a formula of everything that is in the substance; for the universal will be none the less the substance of something; e.g., "man" will be the substance of the man in whom it is present. Thus the same thing will happen again; e.g. "animal" will be the substance of that in which it is present as peculiar to it. [*Metaphysics* Z 13, 1038ᵇ1–34 (*Op. cit.*)]

Uses of species concepts in natural history

Very extensive genera of animals, into which other subdivisions fall, are the following: one, of birds, one, of fishes; and another, of cetaceans. Now all these creatures are blooded.

There is another genus of the hard-shell kind, which is called the shell-fish; and another of the soft-shell kind, not designated by a single term, such as the crayfish and the various kinds of crabs and lobsters; and another of molluscs, as the two kinds [*genē*] of calamary and the cuttle-fish; that of insects is different.
...

Of the other animals the genera are not extensive. For in them one species does not comprehend many species; but in one case, as man, the species is simple, admitting of no differentiation, while other cases admit of differentiation, but the species lack particular definitions.

...

In the genus that combines all viviparous quadrupeds are many species, but under no common appellation. They are only named as it were one by one, as man is—e.g., the lion, the stag, the horse, the dog, and so on; though there is actually a single genus in the case of the so-called bushy-tailed animals, such as the horse, the ass, the mule, the jennet, and the animals that are called mules in Syria,—from their resembling mules, though they are not strictly of the same species, for they mate and breed from one another. [*History of Animals* I.6 490[b7]–491[a]4 (*Op. cit.*)]

For animals that copulate, of one and the same species (*genesi*), the age for maturity is in most species, tolerably uniform, unless it occurs prematurely by reason of abnormality, or is postponed by physical injury. [*History of Animals*, V.14 544[b]22–23 (*Op. cit.*)]

Of sponges there are three species [*genē*]: the first is of porous texture, the second is close-textured, the third, which is nicknamed 'the sponge of Achilles', is exceptionally fine and close-textured and strong. [*Ibid* V.15 548[a]31–548[b]1 (*Op. cit.*)]

Of bees there are various species [*genē*]. The best kind is a little round mottled insect; another is long and resembles a hornet; a third is black and flat-bellied and is nicknamed 'the robber'; a fourth kind is the drone, the largest of all, but stingless and inactive. [*Ibid* V.22 553[b]9–12 (*Op. cit.*)]

Classification by division—dichotomies rejected

Some writers propose to reach the definitions of the ultimate forms of animal life by bipartite division. But this method is often difficult, and often impracticable.

Sometimes the final differentia of the subdivision is sufficient by itself, and the antecedent differentiae are mere surplusage. Thus in the series Footed, Two-footed, Cleft-footed, the last term is all-expressive by itself, and to append the higher terms is only an idle iteration. Again it is not permissible to break up a natural group, Birds for instance, by putting its members under different bifurcations, as is done in the published dichotomies, where some birds are ranked with animals of the water, and others placed in a different class. The group Birds and the group Fishes happen to be named, while other natural groups have no popular names; for instance, the groups that we may call Sanguineous and Bloodless are not known popularly by any designations. If such natural groups are not to be broken up, the method of Dichotomy cannot be employed, for it necessarily involves such breaking up and dislocation. The group of the Many-footed, for instance, would, under this method, have to be dismembered, and some of its kinds distributed among land animals, others among water animals.

Again, privative terms inevitably form one branch of a dichotomous division, as we see in the proposed dichotomies. But privative terms in their character admit of no subdivision. For there can be no specific forms of a negation, of Featherless for instance or of Footless, as there are of Feathered and of Footed. Yet a generic differentia must be subdivisible; for otherwise what is there that makes it generic rather than specific? There are to be found generic, that is, specifically subdivisible, differentiae; Feathered for instance and Footed. For feathers are divisible into Barbed and Unbarbed, and feet into Manycleft, and Twocleft, like those of animals with bifid hoofs, and Uncleft or Undivided, like those of animals of solid hoofs. Now even with differentiae capable of this specific subdivision it is difficult enough so to make the classification that each animal shall be comprehended in one subdivision and in not more than one … ; but far more difficult, impossible, it is to do this if we start with a dichotomy into two contradictories. For each differentia must be presented by some species. There must be some species, therefore, under the privative heading. Now specifically distinct animals cannot present in their substance a common undifferentiated element, but any apparently common element must really be differentiated. (Bird and Man for instance are both Two-footed, but their two-footedness is diverse and differentiated. And if they are sanguineous they must have some difference in their blood, if their blood is part of their substance.) From this it follows that one differentia will belong in two species; and if that is so, it is plain that a privative cannot be a differentia.

Again, if the species are indivisible, and the differentiae are indivisible, and if no differentia be common to several groups, the number of differentiae must be equal to the number of species. If a differentia though not divisible could yet be common to several groups, then it is plain that in virtue of that common differentia specifically distinct animals would fall into the same division. It is necessary then, if the differentiae, under which are ranged all the indivisible groups, are specific characters, that none of them shall be common; for otherwise, as already said, specifically distinct animals will come into one and the same division. But no one indivisible group must be found in more than one division; different groups must not be included in the same division; and every group must be found in some division. It is plain then that we cannot get at the ultimate specific species of the animal, or any other, kingdom by bifurcate division. If we could, the number of ultimate differentiae would equal the number of ultimate animal species. For assume an order of beings whose prime differentiae are White and Non-white. Each of these branches will bifurcate, and their branches again, and so on till we reach the differentiae, whose number will be four or some other power of two, and will also be the number of the ultimate species.

(A species is constituted by the combination of differentia and matter. For no part of an animal is purely material or purely immaterial; nor can a body, independently of its condition, constitute an animal or any of its parts, as has repeatedly been observed.) [*Parts of Animals* I.3 642^{b}22–643^{a}27 (*Op. cit.*)]

Species and spontaneous generation
Again, some creatures come into being neither from parents of the same kind [*genē diaphorōn*] nor from parents of a different kind... [*Generation of Animals* I.18 723^{b}4–5 *Op. cit.*]

The potential eternality of species
Now some existing things are eternal and divine whilst others admit of both existence and non-existence. But that which is noble and divine is always, in virtue of its own nature, the cause of the better in such things as admit of being better or worse, and what is not eternal does admit of existence and non-existence, and can partake in the better and the worse. And soul is better than body, and the living, having soul, is thereby better than the lifeless which has none, and being is better than not being, living than not living. These, then, are the reasons of the generation of animals. For since it is impossible that such a class of things as animals should be of an eternal nature, therefore that which comes into being is eternal in the only way possible. Now it is impossible for it to be eternal as an individual—for the substance of the things that are in is the particular; and if it were such it would be eternal—but it is possible for it as a species. This is why

there is always a class of men and animals and plants. [*Generation of Animals* II.1 731ᵇ24–732ᵃ1 (*Op. cit.*)]

Sterility and hybrids

As a general rule, wild animals are at their wildest in Asia, their boldest in Europe and most diverse in form in Libya [Africa]; in fact, there is an old saying. 'Always something fresh in Libya.'

It would appear that in that country animals of diverse species[6] meet, on account of the rainless climate, at the watering places, and they pair together; and that such pairs breed if they be nearly of the same size and have periods of gestation of the same length. For they are tamed down in their behaviour towards each other by extremity of thirst. ... Elsewhere also offspring are born to heterogeneous pairs; thus in Cyrene the wolf and the bitch will couple and breed; and the Laconian hound is a cross between the fox and the dog. They say that the Indian dog is a cross between the tiger and the bitch, not the first cross, but a cross in the third generation; for they say that the first cross is a savage creature. They take the bitch to a lonely spot and tie her up—and many are eaten, unless the beast is eager to mate. [*History of Animals* VIII.28 606ᵇ16–607ᵃ7 (*Op. cit.*)]

Copulation takes place naturally between animals of the same kind [*homogenesin*]. However, those also unite whose nature is near akin and whose form [*eidei*] is not very different, if their size is much the same and if the periods of gestation are equal. In other animals such cases are rare, but they occur with dogs and foxes and wolves and jackals; the Indian dogs also spring from the union of a dog with some wild dog-like animal. A similar thing has been seen to take place in those birds that are salacious, as partridges and hens. Among birds of prey hawks of different form [*eidei*] are thought to unite, and the same applies to some other birds. ... And the proverb about Libya, that Libya is always producing something new, is said to have originated from animals of different species [*homophulē allēlois*] uniting with one another in that country, for it is said that because of the want of water all meet at the few places where springs are to be found, and that even different kinds unite [*homogenē*] in consequence.

Of the animals that arise from such union all except mules are found to copulate again with each other and to be able to produce young of both sexes, but mules alone are sterile, for they do not generate by union with one another or with other animals. The problem why any individual, whether male or female, is sterile is a general one, for some men and women are sterile, and so are other

---

6   Neither the term *eidos* nor *genos* is used here directly. The Greek is:
     dia gar tēn anonbrian misgesthai dokei anantonta pros ta hudatia kai ta me homophulē

animals in their several kinds, as horses and sheep. But this kind, of mules, is universally so. The causes of sterility in other animals are several. [*Generation of Animals* II.8 746ᵃ29–746ᵇ22 (*Op. cit.*)]

Epicurus of Samos (341–270BCE), and Titus Lucretius Carus (c.94BCE–c.50BCE)

Epicurus' writings, in particular his *On Nature*, have been lost to us, but Lucretius offers what most scholars think is a very faithful recounting of Epicurus' teachings, so I will take his work as dutifully reporting Epicurus' invention of what I call the *Generative Conception of Species*: that natural or living species have a generative order or power that makes progeny resemble children, and development to occur in a repeated orderly sequence.

If things could be created out of nothing, any kind of things could be produced from any source. In the first place, men could spring from the sea, squamous fish from the ground, and birds could be hatched from the sky; cattle and other farm animals, and every kind of wild beast, would bear young of unpredictable species, and would make their home in cultivated and barren parts without discrimination. Moreover, the same fruits would not invariably grow on the same trees, but would change: any tree could bear any fruit. Seeing that there would be no elements with the capacity to generate each kind of thing, how could creatures constantly have a fixed mother? But, as it is, because all are formed from fixed seeds, each is born and issues out into the shores of light only from a source where the right ultimate particles exist. And this explains why all things cannot be produced from all things: any given thing possesses a distinct creative capacity. [*On the Nature of Things* (Lucretius 1969: 38, Book I. *l*155–191)]

Pliny the Elder (Gaius Plinius Secundus, 23–79)

Pliny's Natural History was the standard educated person's encyclopedia of the natural world from its publication to the 18th century. He in particular relies upon Aristotle's *Historia Animalium*, although it is noteworthy that he feels able to disagree with Aristotle when he sees fit. Here he repeats Aristotle's claim that animals—big cats in this case—will interbreed at waterholes in Africa.

The noble appearance of the lion is more especially to be seen in that species which has the neck and shoulders covered with a mane, which is always acquired at the proper age by those produced from a lion; while, on the other hand, those that are the offspring of the pard, are always without this distinction. The female also has no mane. The sexual passions of these animals are very violent, and render the male quite furious. This is especially the case in Africa, where, in consequence of the great scarcity of water, the wild beasts assemble in great numbers on the banks of a few rivers. This is also the reason why so many curious varieties of animals are produced there, the males and females of various species coupling promiscuously with each other. Hence arose the saying, which was common in Greece even, that "Africa is always producing something new." The lion recognizes, by the peculiar odour of the pard, when the lioness has been unfaithful to him, and avenges himself with the greatest fury. Hence it is, that the female, when she has been guilty of a lapse, washes herself, or else follows the lion at a considerable distance. I find that it was a common belief, that the lioness is able to bear young no more than once, because, while delivering herself, she tears her womb with her claws. Aristotle, however, gives a different account; a man of whom I think that I ought here to make some further mention, seeing that upon these subjects, I intend, in a great measure, to make him my guide. Alexander the Great, being inflamed with a strong desire to become acquainted with the natures of animals, entrusted the prosecution of this design to Aristotle, a man who held the highest rank in every branch of learning; for which purpose he placed under his command some thousands of men in every region of Asia and Greece, and comprising all those who followed the business of hunting, fowling, or fishing, or who had the care of parks, herds of cattle, the breeding of bees, fish-ponds, and aviaries, in order that no creature that was known to exist might escape his notice. By means of the information which he obtained from these persons, he was enabled to compose some fifty volumes, which are deservedly esteemed, on the subject of animals; of these I purpose to give an epitome, together with other facts with which Aristotle was unacquainted; and I beg the kind indulgence of my readers in their estimate of this work of mine, as by my aid they hastily travel through all the works of nature, and through the midst of subjects with which that most famous of all kings so ardently desired to be acquainted. [*Natural History* VIII 17.42 (Pliny 1855)]

Porphyry the Phoenician (c.232–c.305)

Porphyry's neo-Platonic conflation of Aristotle's immanent forms, and Plato's transcendent ones (Boodin 1943), is the mediate source of the *diairesis* to the western medieval tradition, via Boethius, and triggered off the debates over nominalism. But here it is interesting that for Porphyry, the *shape* of something is its form, a geometrical slant not usually found.

We call a species, first, the shape of anything … We also call a species what is under the genus of the sort presented … [*Isagoge*, §2 (Porphyry and Barnes 2003: 5) c.270]

Anicius Manlius Severinus Boethius (480–524 or 525)

Boethius is better known for his *Consolations of Philosophy*, but much more significant in subsequent philosophy was his translation and commentary on Porphyry's *Isagoge* (*Introduction*). The following passage kicked off the nominalist tradition.

As for genera and species, [Porphyry] says, I shall decline for the present to say (1) whether they subsist or are posited in bare [acts of] understanding only, (2) whether, if they subsist, they are corporeal or incorporeal, and (3) whether [they are] separated from sensibles or posited in sensibles and agree with them. For that is a most noble matter, and requires a longer investigation. [*Anicii Manlii Severini Boethii In Isagogen Porphyrii commenta*, editio 2a, lib. I, ca. 10–11, Samuel Brandt, ed., p. 159 line 3—p. 167 line 20. Translation by Paul Spade, unpublished (used with permission)]

Aurelius Augustinus, Augustine of Hippo (354–430)

Augustine, like a number of classical authors, has been claimed as a creationist as well as an evolutionist, with respect to species. These categories, which apply much later, are anachronistic. Augustine was neither, but he thought God was the source of all forms. Forms might change, but not God. [This is mirrored in the almost exclusive use during the 19th century Catholic revival of Aristotle and Thomas of the term "essence" to apply to God's nature, not natural species.] Augustine is sometimes claimed as a mutabilist with respect to species, but this is clearly not the case, as seen by the second quotation on the creation of plants. He thinks that God created plants by making them potentially in the soil. This is not mutabilism—at best it is spontaneous generation. See below for Aquinas' agreement on this.

… all forms of mutable things, whereby they are what they are (of what nature soever they be) have their origin from none but Him that is true and unchangeable. Consequently, neither the body of this universe, the figures, qualities, motions, and elements, nor the bodies in them from heaven to earth, either vegetative as trees, or sensitive also as beasts, or reasonable also as men, nor those that need no nutriment but subsist by themselves as the angels, can have being but from Him who has only simple being. [*Civitas Dei* (*The city of God*) Book VIII, chapter 6]

Where, then, were they [*plants, when they were created*]? Were they in the earth in the "reasons" or causes from which they would spring, as all things already exist in their seeds before they evolve [*develop—JSW*] in one form or another and grow into their proper kinds in the course of time? ... it appears [*from Scripture—JSW*] ... that the seeds sprang from the crops and trees, and that the crops and trees themselves came forth not from seeds but from the earth. [*De Genesi Ad Litteram*, (*The literal meaning of Genesis*) c. 390 AD, Book V, chapt 4 (Augustine 1982: 151f)]

Peter Abelard (1079–1142)

Abelard is here rejecting the idea that universal terms are aggregates of individual objects.

... clearly the species or genus is not a multitude of discrete things ... [(McKeon 1929: 257)]

Peter of Lombard (c.1100–1160)

Lombard makes what came to be a crucial distinction, between the thing and the appearance or "sign" of the thing. Later this becomes known as "Peirce's triad", with the addition of what is implicit here, or the thought.

A sign, however, is the thing beside the species, which it bears upon the senses, causing something else out of itself to come into (one's) thinking. [*The Sentences*, Book IV, Pt 1, Chapter 3 (from the Franciscan Archive, <http://www.franciscan-archive.org/lombardus/index.html#writings>, accessed 3 June 2005)]

Frederick II of Hohenstaufen (1194–1250)

Frederick is a real character. A Norman king and Holy Roman Emperor with a passion for falconry (he spent as much time on falconry as he did on affairs of state, and reportedly told the Khan he would submit if he could become the Khan's falconer), he was forced to write his own manual on the "sport", called *The Art of Hunting with Birds*. In this he demonstrated a surprisingly modern view of species not marked by external form or colouration but by ability to interbreed (Wilkins 2005), and he also demonstrated scepticism about the well-known tale of the Barnacle Goose spring from barnacles, which themselves spontaneously generated out of rotting driftwood. His comments on Aristotle indicate that Aristotle was not always an authority for him, and in falconry not at all. So far from being a fixist about species, it appears he thought Nature could modify organisms to suit their conditions.

> *Inter alia*, we discovered by hard-won experience that the deductions of Aristotle, whom we followed when they appealed to our reason, were not entirely to be relied upon, more particularly in his descriptions of the characters of certain birds.

> There is another reason why we do not follow implicitly the Prince of Philosophers: he was ignorant of the practice of falconry—an art which to us has ever been a pleasing occupation, and with the details of which we are well acquainted. In his work "Liber Animalium" we find many quotations from other authors whose statements he did not verify and who, in their turn, were not speaking from experience. Entire conviction of the truth never follows mere hearsay. (Wood and Fyfe 1943: 3f)

> … a description of the essential characters of individual birds [i.e., of a species] is more difficult to furnish, whether they resemble or are different from another in the shape of the limbs, the movements they make, the way they feed, the care of their young, their mode of flight, and their style of defense. Let it, however, be remembered that, in general, their bodily conditions and their other peculiarities are due to definite causes. [p10]

> Nature in her endeavor to preserve the race by the continuous multiplication of individuals has decreed that every species of the animal kingdom, whether it progresses by the use of wings or walks on the ground, shall take pleasure in sexual union so that they may seek instinctively to bring about such enjoyment. [p49]

> [It] must be held, then, that for each species, and each individual of the species, Nature has provided and made, of convenient, suitable, material, organs adapted to individual requirements. By means of these organs the individual has perfected the functions needful for himself. It follows, also, that each individual, in accordance with the particular form of his organs and the characteristics inherent in them, seeks to perform by means of each organ whatever task is most suitable to the form of that organ. [p57]

Albertus Magnus (Albert of Lauingen, 1193–1280)

Albert the Great was also able to make use of Frederick's extensive falconries, and wrote an extensive volume in his multi-volume *De Animalibus* on birds. In this he showed, too, scepticism about the spontaneous generation claim for the Barnacle Goose by actually breeding them and producing eggs and goslings in the natural way.

For Albert, species are maintained by sexual reproduction. Note Stannard's comment about the use of *genus, species, forma* and other terms: "In medieval nomenclature, *genus, species, varietas,* and *forma* were used interchangeably. Sometimes Albert uses *genus* to denote what we would accept as a *genus* ... but in other places his *genus* is practicably equivalent to our *species* ..." (Stannard 1980a: 366n).

> ... the Creator unmistakenly wanted the animal kingdom to endure as a stable entity, never to die out. To this end He ordained the class of animals to be continually renewed by the coupling of the sexes and reproduction of the species so that none would be lost. [*De animalibus,* book 22, citing *De coito* of Consantine of Cassino (Constantinus Africanus, c.1010–1087) (Albertus Magnus 1987: 59)]

Thomas Aquinas (c.1225–1274)

Thomas, a student of Albert's is credited, somewhat incorrectly (there were precursors), with introducing Aristotle into Christian thought. However, it cannot be denied that he was very influential in having Aristotle's logic accepted, and becoming, 600 years after his death, the "official" philosopher of the Catholic Church.

Thomas distinguishes between material (substantive) division and formal (logical) division. He notes that Augustine allowed that individual species of plants might be generated from the potential placed there by God at creation. For animals, though, he is less flexible, and appears to think that animal species are relatively fixed even if they can occasionally hybridise with some success.

> These [*species infimae* or *specialissimae*] are called individuals, in so far as they are not further divisible formally. Individuals however are called particulars in so far as they are not further divisible neither materially nor formally. [*In lib. X Met.* Lect 10, 2123 (quoted in McKeon 1930: 498)

> But concerning the production of plants, Augustine's opinion differs from that of others. For other commentators, in accordance with the surface meaning of the text, consider that the plants were produced in act in their various species on this third day; whereas Augustine (*Gen. ad lit.* v, 5; viii, 3) says that the earth is said to have then produced plants and trees in their causes, that is, it received then the power to produce them. He supports this view by the authority of Scripture, for it is said (Gn. 2:4,5): "These are the generations of the heaven and the earth, when they were created, in the day that ... God made the heaven and the earth, and every plant of the field before it sprung up in the earth, and every herb of the ground before it grew." Therefore, the production of plants in their causes, within the earth, took place before they sprang up from the earth's surface. And this is

confirmed by reason, as follows. In these first days God created all things in their origin or causes, and from this work He subsequently rested. Yet afterwards, by governing His creatures, in the work of propagation, "He worketh until now." Now the production of plants from out the earth is a work of propagation, and therefore they were not produced in act on the third day, but in their causes only. However, in accordance with other writers, it may be said that the first constitution of species belongs to the work of the six days, but the reproduction among them of like from like, to the government of the universe. And Scripture indicates this in the words, "before it sprung up in the earth," and "before it grew," that is, before like was produced from like; just as now happens in the natural course by the production of seed. Wherefore Scripture says pointedly (Gn. 1:11): "Let the earth bring forth the green herb, and such as may seed," as indicating the production of perfection of perfect species, from which the seed of others should arise. Nor does the question where the seminal power may reside, whether in root, stem, or fruit, affect the argument. [*Summa* I.69.2. Obj 3 (Aquinas 1947)]

Species, also, that are new, if any such appear, existed beforehand in various active powers; so that animals, and perhaps even new species of animals, are produced by putrefaction by the power which the stars and elements received at the beginning. Again, animals of new kinds arise occasionally from the connection of individuals belonging to different species, as the mule is the offspring of an ass and a mare; but even these existed previously in their causes, in the works of the six days. ... Hence it is written (Eccles. 1:10), "Nothing under the sun is new, for it hath already gone before, in the ages that were before us." [*Summa* I.73.1. Obj 3. Reply to Objection 3 (*Op. cit.*)]

William of Ockham or Occam, (c1300–1349)

William of Ockham is perhaps the most enduringly influential of the medieval nominalists. For him, logical species are just a way to recollect similar individuals already encountered, and a general term cannot be abstracted from a single individual, but only a number of individuals encountered, as he says in *The seven quodlibeta*, quod.1, ques. 13 (in McKeon 1930: 365). Hence to assert that something coming from a distance is an animal, one must already have that concept by recollecting prior individual animals. This demonstrates that the nominalist is attempting a "bottom-up" form of classification, based on observed cases. Moreover, species, logical or otherwise, are things which have the same "power" (quod.V, ques. 2), but concepts of them are of the "second intention" (that is, have been categorized by the mind):

Ockham's influence on subsequent philosophy, theology and science has been enormous. Here, he takes from Avicenna a distinction of intentions—a concept of the first intention is an idea formed by interaction with things, which he thinks are natural. A concept of the second intention signifies facts (which are those of the Aristotelian logic) about those of the first intention. So to say something is a species is in effect to say it has a general name.

> … that concept is called a second intention which signifies precisely intentions
> naturally significative, or which sort are *genus, species, difference,* and others of
> this sort [i.e., heads of predicables] for as the concept of man is predicated of all
> men …, so too one common concept, which is the second intention, is predicated
> of first intentions [a conception of a thing formed by the first or direct application
> of the mind to the individual object; an idea or image] which stand for things, as
> in saying, *man is a species, ass is a species, whiteness is a species, animal is a genus,*
> *body is a genus*; in the manner in which *name* is predicated of different names
> … and this second intention thus signifies first intentions naturally, and can
> stand for them in a proposition, just as the first intention signifies external things
> naturally. [*The seven quodlibeta,* quod IV ques. 19 (in McKeon 1930: 388)]

Nicholas of Cusa (1401–1464)

Cusa is a neo-Platonist, influenced by the recent translation into Latin of the works of Plotinus and Pophyry, among others. Accepting Aristotle's Ten Topics ("Ten Words"; or most general categories from which all others can be divided), he instead says that a species is a "contraction" of the broader class into a single thing, and that the universe is contracted in each of its parts (the doctrine of the macrocosm/microcosm). Again he asserts that members of species have a "common nature" which exists whether there are forms or not. Because this is a contraction, individuals can have varying degrees of it, and so he is not an essentialist in the usual manner.

> … the universe is contracted in each particular through three grades. Therefore,
> the universe is, as it were, all of the ten categories *[generalissima]*, then the genera,
> and then the species. And so, these are universal according to their respective
> degrees; they exist with degrees and prior, by a certain order of nature, to the
> thing which actually contracts them. And since the universe is contracted, it is
> not found except as unfolded in genera; and genera are found only in species"
> [*On learned ignorance* §124 (c1440, cf. Hopkins 1981)]

> For example, dogs and the other animals of the same species are united by virtue
> of the common specific nature which is in them. This nature would be contracted
> in them even if Plato's intellect had not, from a comparison of likenesses, formed

for itself a species. Therefore, with respect to its own operation, understanding follows being and living; for [merely] through its own operation understanding can bestow neither being nor living nor understanding" [§126]

Now, the many things in which the universe is actually contracted cannot at all agree in supreme equality; for then they would cease being many. Therefore, it is necessary that all things differ from one another—either *(1)* in genus, species, and number or *(2)* in species and number or *(3)* in number-so that each thing exists in its own number, weight, and measure. Hence, all things are distinguished from one another by degrees, so that no thing coincides with another. Accordingly, no contracted thing can participate precisely in the degree of contraction of another thing, so that, necessarily, any given thing is comparatively greater or lesser than any other given thing. Therefore, all contracted things exist between a maximum and a minimum, so that there can be posited a greater and a lesser degree of contraction than [that of] any given thing. [§182, p124]

Therefore, no species descends to the point that it is the minimum species of some genus, for before it reaches the minimum it is changed into another species; and a similar thing holds true of the [would-be] maximum species, which is changed into another species before it becomes a maximum species. ... Accordingly, it is evident that species are like a number series which progresses sequentially and which, necessarily, is finite, so that there is order, harmony, and proportion in diversity... [§§ 185–187]

Marsilio Ficino (1433–1499)

Ficino, a little later than Cusa, talks about "natural species" with their own motive forces.

The motion of each of all the natural species proceeds according to a certain principle. Different species are moved in different ways, and each species always preserves the same course in its motion so that it proceeds from this place to that place and, in turn, recedes from the latter to the former, in a certain most harmonious manner. [*Five Questions Concerning the Mind* (Cassirer, Kristeller, and Randall 1948: 194)]

In this period, several different traditions are in play. One is the ongoing logical tradition, which morphs into what has come to be known as the Universal Language Project (Slaughter 1982). This is founded not so much on Aristotle alone, as on the revival of the Neo-Platonism of Plotinus and Porphyry, with its curious mix of Plato and Aristotle. Under this project, the idea was to generate a universal set of semantic or logical categories, under which all things must exhaustively and exclusively be classed. At the same time, the traditional logic, now called "scholastic logic", continued to employ increasingly sophisticated versions of the *diairesis*. Both of these traditions were engaged in metaphysical argument, and so the categories were thought to picture the way the world is. This was attacked by Locke, who instead proposed a nominalist view of general categories.

Almost independently, though, the science of natural history began, after the Reformation and Counter-Reformation had settled down, and people such as Bauhin and Cesalpino began to catalogue and attempt to make sense of the plethora of folk and herbal descriptions of types of organisms, particularly plants. This almost immediately led to questions of how to arrange species, and how to name them. The first exhaustive *Flora* of a region was produced by John Ray, for Cambridgeshire, and it is he who first defined a purely "natural" conception of species. He also produced the first fixism of species as well, on pietistic grounds rather than due to any logical considerations.

Ray's definition and fixism became, via Linnaeus, the "standard" view of the times, but it is worth observing that not long after Linnaeus began to define species as fixed created kinds, Pierre Maupertuis introduced the first explicit transformism of species. In other words, by setting up the contrast of species as fixed or mutable, Ray and Linaneus made it a question whether they, in fact, were. Linnaeus did not argue for the standard view of species, but merely asserted it in aphorisms, but later in life, and influentially upon the botanical community, he argued that hybrids could form novel species, and became a limited transformist himself.

Since Aristotle, naturalists had known that deviations from the type could occur, even in properties that were diagnostically essential to the species. A Platonist (or neo-Platonist) would have no trouble here, for the type would be self-instantiating, and individual instances that were deviant would not affect the type itself, but for Aristotelians, for whom the type just *was* the groups of individual organisms, this was a real problem. Hence, the persistence of a generative conception, in which there is a causal type that stabilises organisms around the type, enabled them to treat deviation as deviance. This is clear in, say Grew's comments, in which the "essence" of the plant is not the essence of a *species*, but of the constitution that makes the individual plants come out the same in reproduction and growth.

Peter Ramus (1515–1572)

Ramus introduced the notion of "Porphyry's tree", a diagram that expressed the logical dichotomy from the Topic to the infimae species, as a tree. It is also sometimes called "Ramus' Tree". It is clear from this excerpt that species and genus are semantic categories: answers to questions in an erotetic logic, and not necessarily something that has causal power.

> A species is the thing itself concerning which the genus answers [in replying to the question, What is it?]: thus the genus man answers concerning Plato, the genus dialectician concerning this particular dialectician. [*Training in Dialectic*, 1543, fol. 14 (quoted in Ong 1958: 204)]

Francis Bacon (1561–1626)

Bacon is famously the founder of "induction" (in a very primitive and unsophisticated manner) but he also famously decried final causes as "barren virgins" in sciences not involving human beings, and here treats general concepts (universals) as abstractions from the senses, and which may be subject to variation and deviation.

He was involved in what came to be known as the Universal Language Project, in which an aim (*the* aim?) of science was to come up with a linguistic scheme that would identify all natural kinds of objects in the world.

Bacon's discussion of transmutation of species appears to some—e.g., Samuel Butler (1879)—to make him an evolutionist, but it is not clear that he is here referring to biological species so much as the ordinary sense of "kind" or "form", as witness his treating the metamorphosis of insects from caterpillars to "flies" as a specific transmigration. His discussion of the reversion of cultivars into wild forms is loose enough to indicate that he simply lacked a biological notion of species entirely, as we might expect at this period.

Our notions of less general species, as Man, Dog, Dove and of the immediate perceptions of the sense, as Hot, Cold, Black, White, do not materially mislead us; yet even these are sometimes confused by the flux and alteration of matter and mixing of one thing with another. All the others which men have adopted are but wanderings, not being abstracted and formed by proper methods. [*Novum Organum* I.xvi]

Among Prerogative Instances I will put in eighth place *Deviating Instances*, that is, errors, vagaries, and prodigies of nature, wherein nature deviates and turns aside from her ordinary course. Errors of nature differ from singular instances in this, that the latter are prodigies of species, the former of individuals. Their use is pretty much the same, for they correct the erroneous impressions suggested to the understanding by ordinary phenomena, and reveal common forms. ...

... we have to make a collection or particular natural history of all prodigies and monstrous births of nature; of everything in short that is in nature new, rare, and unusual. [II.xxix]

... I will put in the ninth place *Bordering Instances*, which I will also call *Participles*. They are those which exhibit species of bodies which seem to be composed of two species, or to be rudiments between one species and another. ...

Examples of these are: moss, which holds a place between putrescence and a plant; some comets, between stars and fiery meteors; flying fish, between birds and fish; bats, between birds and quadrupeds; also the ape, between man and beast—

Simia quam similes turpissima bestia nobis;[1]

Likewise the biformed births of animals, mixed of different species, and the like. [*Novum Organon.* 1605 (Bacon 1960: 178f)]

Nature exists in three states, and is subject, as it were, to three kinds of regimen. Either she is free and develops herself in her own ordinary course, or she is forced out of her proper state by the perverseness and insubordination of matter and the violence of impediments, or she is constrained and molded by art and human ministry. The first state refers to the "species" of things; the second

---

1   "How like us is that very ugly beast, the monkey", Ennius (239-169?BCE) as quoted in Cicero's "On the Nature of the Gods". Thanks to Tom Scharle for the reference and translation.

to "monsters"; the third to "things artificial." [*Preparative Toward Natural and Experimental History* 1620 (Bacon 1863), aphorism I]

But the more difficult and laborious the work is, the more ought it to be discharged of matters superfluous. And therefore there are three things upon which men should be warned to be sparing of their labor, as those which will immensely increase the mass of the work and add little or nothing to its worth.

...

Secondly, that superfluity of natural histories in descriptions and pictures of species, and the curious variety of the same, is not much to the purpose. For small varieties of this kind are only a kind of sports and wanton freaks of nature and come near to the nature of individuals. They afford a pleasant recreation in wandering among them and looking at them as objects in themselves, but the information they yield to the sciences is slight and almost superfluous. [*Ibid* Aphorism III]

For the nature of things is so distributed that the quantity or mass of some bodies in the universe is very great, because their configurations require a texture of matter easy and obvious, such as are those four bodies which I have mentioned; while of certain other bodies the quantity is small and weakly supplied, because the texture of matter which they require is very complex and subtle, and for the most part determinate and organic, such as are the species of natural things—metals, plants, animals. [*Ibid* Aphorism IV]

Next come Histories of Species.

26. History of perfect Metals, Gold, Silver; and of the Mines, Veins, Marcasites of the same; also of the Working in the Mines.

27. History of Quicksilver.

28. History of Fossils; as Vitriol, Sulphur, etc.

29. History of Gems; as the Diamond, the Ruby, etc.

30. History of Stones; as Marble, Touchstone, Flint, etc.

31. History of the Magnet.

32. History of Miscellaneous Bodies, which are neither entirely Fossil nor Vegetable; as Salts, Amber, Ambergris, etc.

33. Chemical History of Metals and Minerals.

34. History of Plants, Trees, Shrubs, Herbs; and of their parts, Roots, Stalks, Wood, Leaves, Flowers, Fruits, Seeds, Gums, etc.

35. Chemical History of Vegetables.

36. History of Fishes, and the Parts and Generation of them.

37. History of Birds, and the Parts and Generation of them.

38. History of Quadrupeds, and the Parts and Generation of them.

39. History of Serpents, Worms, Flies, and other insects; and of the Parts and Generation of them.

40. Chemical History of the things which are taken by Animals. [a list of topics of science, *Ibid* Aphorism X]

518. The rule is certain, that plants for want of culture degenerate to be baser in the same kind; sometimes so far as to change into another kind. 1. The standing long, and not being removed, maketh them degenerate. 2. Drought, unless the earth itself be moist, doth the like. 3. So doth being removed into worse earth, or forebearing to compost the earth; as we see that water-mint turneth into field-mint, and the cole-wort into rape, by neglect, &c. [*Sylva Sylvarum* 1627 (Bacon 1863), Century VI]

This work of the transmutation of plants one into another, is *inter magnalia naturae*: for the transmutation of species is, in the vulgar philosophy, pronounced impossible; and certainly it is a thing of difficulty, and requireth deep search into nature; but seeing there appear some manifest instances of it, the opinion of impossibility is to be rejected, and the means therefore to be found out. We see that in living creatures that come of putrefaction, there is much transmutation of one into another; as caterpillars turn into flies, &c. And it should seem probab;e that whatsoever creature, having life, is generated without seed, that creature will change out of one species into another. For it is the seed, and the nature of it, which locketh and boundeth in the creature, that it doth not expatiate. Se we may well conclude, that seeing the earth of itself doth put forth plants without seed, therefore plants may well have transmigration of species. Wherefore, wanting instances which do occur, we shall give directions of the most likely trials… [*Ibid* Aphorism 525]

John Locke (1632–1704)

Locke follows after the effective failure of the Universal Language Project, and takes nominalism to its logical conclusion. All universals are merely matter of convenience and communication, including (logical) species. And even they are subject to variation (such as when we ask if a deformed human truly is human enough to baptise). That there are real essences, or causally necessary properties of things, he doesn't doubt, but he does think we have no access to them, only to their "nominal essences", and so our general notions, such as "species", are merely conveniences of little scientific import.

He makes the point that the Latin terms have no authority that good English terms like "sort" or "kind" do not. It is clear Locke is not even a logical essentialist. Since he saw what he did as "clearing the undergrowth" for science, this is rather significant in the development of the species concept. He shortly preceded, and knew, the works of John Ray (below) and Ray personally, and was influential on Buffon and Lamarck.

For the natural tendency of the mind being towards knowledge; and finding that, if it should proceed by and dwell upon only particular things, its progress would be very slow, and its work endless; therefore, to shorten its way to knowledge, and make each perception more comprehensive, the first thing it does, as the foundation of the easier enlarging its knowledge, either by contemplation of the things themselves that it would know, or conference with others about them, is to bind them into bundles, and rank them so into sorts, that what knowledge it gets of any of them it may thereby with assurance extend to all of that sort; and so advance by larger steps in that which is its great business, knowledge. This, as I have elsewhere shown, is the reason why we collect things under comprehensive ideas, with names annexed to them, into genera and species; i.e. into kinds and sorts. [*Essay on human understanding*, Bk II, chap. 32, §6]

The learning and disputes of the schools having been much busied about genus and species, the word essence has almost lost its primary signification: and, instead of the real constitution of things, has been almost wholly applied to the artificial constitution of genus and species. It is true, there is ordinarily supposed a real constitution of the sorts of things; and it is past doubt there must be some real constitution, on which any collection of simple ideas co-existing must depend. But, it being evident that things are ranked under names into sorts or species, only as they agree to certain abstract ideas, to which we have annexed those names, the essence of each genus, or sort, comes to be nothing but that abstract idea which the general, or sortal (if I may have leave so to call it from sort, as I do general from genus), name stands for. And this we shall find to be that which the word essence imports in its most familiar use. [Bk III, chap. III, §15]

This shows Species to be made for Communication.—The reason why I take so particular notice of this is, that we may not be mistaken about *genera* and *species*, and their *essences*, as if they were things regularly and constantly made by nature, and had a real existence in things; when they appear, upon a more wary survey, to be nothing else but an artifice of the understanding, for the easier signifying such collections of *ideas* as it should often have occasion to communicate by one general term; under which divers particulars, as far forth as they agreed to that abstract *idea*, might be comprehended. And if the doubtful signification of the word *species* may make it sound harsh to some, that I say the species of mixed modes are "made by the understanding"; yet, I think, it can by nobody be denied that it is the mind makes those abstract complex *ideas* to which specific names are given. And if it be true, as it is, that the mind makes the patterns for sorting and naming of things, I leave it to be considered who makes the boundaries of the sort or *species*; since with me *species* and *sort* have no other difference than that of a Latin and English *idiom*. [Bk III, chap. V, §9]

But supposing that the *real essences* of substances were discoverable by those that would severely apply themselves to that inquiry, yet we could not reasonably think that the *ranking of things under general names was regulated by* those internal real constitutions, or anything else but *their obvious appearances*; since languages, in all countries, have been established long before sciences. So that they have not been philosophers or logicians, or such who have troubled themselves about *forms* and *essences*, that have made the general names that are in use amongst the several nations of men: but those more or less comprehensive terms have, for the most part, in all languages, received their birth and signification from ignorant and illiterate people, who sorted and denominated things by those sensible qualities they found in them; thereby to signify them, when absent, to others, whether they had an occasion to mention a sort or a particular thing. [Bk III, chap. 6, §25]

Gottfried Wilhelm Leibniz (also Leibnitz or von Leibniz, 1646–1716)

Leibniz was immensely impressed and stimulated by Locke, but, as we see here, he was more optimistic about the reach of science. Unfortunately, Locke died before Leibniz completed this book, so it was not published in his lifetime, out of respect for an opponent who could no longer defend his views.

PHIL. §25. Languages were established before sciences, and things were put into species by ignorant and illiterate people. [Representing Locke, above]

THEO. This is true, but the people who study a subject-matter correct popular notions. Assayers have found precise methods for identifying and separating metals, botanists have marvelously extended our knowledge of plants, and experiments have been made on insects that have given us new routes into the knowledge of animals. However, we are still far short of halfway along our journey. [Representing Leibniz' view, from *New Essays on Human Understanding* (Leibniz 1996: 319)]

Andreas Cesalpino (1519–1603)

Cesalpino is a very early naturalist. Here, he appeals to a "law of nature" not a logical principle, to define species. But he refers to "essences" (although note that the word for "substance" is use here, not "essentia") for species in contrast to "accidents", so he may be taken as an essentialist about natural species.

> That according to the law of nature like always produces like and that which is of the same species with itself. [*Quaestionum peripateticarum, libri V*, chapter 13 (quoted in Sachs 1890: 52)
>
> We seek similarities and dissimilarities of form, in which the essence ('substantia') of plants consists, but not of things which are merely accidents of them ('quae accidunt ipsis'). [chapter 14, quoted in *loc. cit.*]
>
> Since science consists in grouping together of like and the distinction of unlike things, and since this amounts to the division into genera and species, that is, into classes based on characters (*differentiae*) which describe the fundamental nature of the things classified, I have tried to do this in my general history of plants, … [*De plantis* (translated in Morton 1981: 135)]

Robert Hooke (1635–1703)

Hooke, a polymath known best for disputing whether he or Newton had discovered the inverse square law of gravity, published *Micrographia* in 1665, which was widely received. In it he uses microscopes to investigate small life forms and the structure of tissues, being the first to name the cells of plants. Like Grew (below), he thought that he was uncovering the essences of the organisms. Here he discusses whether plant species need to form from seed, suggesting that this was a key property of species. He decides not, but only because the deity is free to form species however he likes, whether directly or through the action of secondary causes.

> But to refer this Discourse of Animals to their proper places, I shall add, that though one should suppose, or it should be prov'd by Observations; that several of these kinds of Plants are accidentally produc'd by a casual *putrifaction*, I see not any great reason to question, but that, notwithstanding its own production was as 'twere casual, yet it may germinate and produce seed, and by it propagate its own, that is, a new Species. For we do not know, but that the Omnipotent and All-wise Creator might as directly design the structure of such a Vegetable, or such an Animal to be produc'd out of such or such a *putrifaction* or change of this or that body, towards the constitution or structure of which, he knew it necessary, or thought it fit to make it an ingredient; as that the digestion or moder-

ate heating of an Egg, either by the Female, or the Sun, or the heat of the Fire, or the like, should produce this or that Bird; or that *Putrifactive* and warm steams should, out of the blowings, as they call them, that is, the Eggs of a Flie, produce a living Magot, and that, by degrees, be turn'd into an *Aurelia*, and that, by a longer and a proportion'd heat, be transmuted into a Fly. Nor need we therefore to suppose it the more imperfect in its kind, then the more compounded Vegetable or Animal of which it is a part; for he might as compleatly furnish it with all kinds of contrivances necessary for its own existence, and the propagation of its own Species, and yet make it a part of a more compounded body: as a Clock-maker might make a Set of Chimes to be a part of a Clock, and yet, when the watch part or striking part are taken away, and the hindrances of its motion remov'd, this chiming part may go as accurately, and strike its tune as exactly, as if it were still a part of the compounded Automaton. So, though the original cause, or seminal principle from which this minute Plant on Rose leaves did spring; were, before the corruption caus'd by the Mill-dew, a component part of the leaf on which it grew, and did serve as a *coagent* in the production and constitution of it, yet might it be so consummate, as to produce a seed which might have a power of propagating the same species: the works of the Creator seeming of such an excellency, that though they are unable to help to the perfecting of the more compounded existence of the greater Plant or Animal, they may have notwithstanding an ability of acting singly upon their own internal principle, so as to produce a Vegetable body, though of a less compounded nature, and to proceed so farr in the method of other Vegetables, as to bear flowers and seeds, which may be capabale of propagating the like. So that the little cases which appear to grow on the top of the slender stalks, may, for ought I know, though I should suppose them to spring from the perverting of the usual course of the parent Vegetable, contain a seed, which, being scatter'd on other leaves of the same Plant, may produce a Plant of much the same kind. [Observation XIX, *Micrographia,* (Hooke 1665)]

John Wilkins (1614–1672)

Bishop John Wilkins was cofounder of and first secretary of, the Royal Society. The Society published his work *Essay Towards a Real Character, and a Philosophical Language* (Wilkins 1668), in which he tried to logically define, and assign a unique word to, all logical species. In the course of this, he employed John Ray (below) to draw up a number of tables of species of animals and plants within his tripartite logic of division. So far as is known, apart from the negative influence on Ray, the sole legacy of the *Essay* is to act as the basis for *Roget's Thesaurus.* However, in the course of this work he does use the term "species" for animals, especially when discussing the size of Noah's Ark, based on the tract *de Arca Noe* by Johannes Buteo (Wood 2007). One interesting point is that he treats the mule as a "mungrel". In an accompanying table, he lists the following "species" on the Ark (by implication, the totality of land species):

Beasts feeding on Hay: Horse, Asse, Camel, Elephant, Bull, Urus, Bisons, Bonasus, Buffalo, Sheep, Stepciseros, Broad-tail, Goat, Stone-buck, Shamois, Antilope, Elke, Hart, Buck, Rein-deer, Roe, Rhinocerot, Camelopard, Hare, Rabbet, Marmotto.

Beasts feeding on Fruits, Roots and Insects: Hog, Baboon, Ape, Monky, Sloth, Porcupine, Hedghog, Squirril, Ginny pig, Ant-bear, Armadilla, Tortoise.

Carnivorous Beasts: Lion, Beare, Tigre, Pard, Ounce, Cat, Civet-cat, Ferret, Polecat, Martin, Stoat, Weesle, Castor, Otter, Dog, Wolf, Fox, Badger, Jackall, Caraguya.

This seeming difficulty is much better solved by *Joh. Buteo* in the Tract *de Arca Noe*, wherein supposing the cubit to be the same with what we now call a foot and a half, he proves Mathematically that there was a sufficient capacity in the *Ark*, for the conteining all those things it was designed for. But because there are some things liable to exception in the Philosophical part of that discourse, particularly in his enumeration of the species of Animals, several of which are fabulous, some not distinct species, others that are true species being left out; therefore I conceive it may not be improper in this place to offer another account of those things.

It is plain in the description which *Moses* gives of the *Ark*, that it was divided into three stories, each of them of ten cubits or fifteen foot high, besides one cubit allowed for the declivity of the roof in the upper story. And 'tis agreed upon as most probable, that the lower story was assigned to contein all the species of beasts, the middle story for their food, and the upper story, in one part of it, for the birds and their food, and the other part for *Noah*, his family and utensils.

In this enumeration I do not mention the Mule, because 'tis a mungrel pro-duction, and not to be rekoned as a distinct species. And tho it be most probable, that the several varieties of Beeves, namely that which is stiled *Urus* [aurochs], *Bisons*, *Bonasus* and *Buffalo*, and those other varieties reckoned under *Sheep* and *Goats*, be not distinct species from *Bull*, *Sheep*, and *Goat*; There being much less difference betwixt these, then there is betwixt several Dogs: And it being known by experience, what various changes are frequently occasioned in the same spe-cies by several countries, diets, and other accidents: Yet I have *ex abundanti* to prevent all cavilling, allowed them to be distinct species, and each of them to be clean Beasts, and consequently such as were to be received in by sevens. As for the *Morse* [walrus], *Seale, Turtle,* or *Sea-Tortoise, Crocodile, Senembi* [iguana], These are usually described to be such kind of *Animals* as can abide in the water, and therefore I have not taken them into the Ark, tho if that were necessary, there would be room enough for them, as will shortly appear. The *Serpentine-kind, Snake, Viper, Slow-worm, Lizard, Frog, Toad,* might have sufficient space for their reception, and for their nourishment, in the Drein or Sink of the *Ark,* which was probably three or four foot under the floor for the standings of the Beasts. As for those lesser Beasts, *Rat, Mouse, Mole,* as likewise for the several species of Insects, there can be no reason to question, but that these may find sufficient room in several parts of the *Ark,* without having any particular Stalls appointed for them. [*Essay (*Wilkins 1668:167f)]

John Ray (1627–1705)

Ray was employed to contribute tables of plant species, and his contributor Francis Willughby animal species, to Bp John Wilkins' *Essay,* which he found mightily constraining and criticisms of that works stung him. The *Essay* was the final flowering of the Universal Language Project, and as such, it marks the beginnings of empirical classification. Ray's subjects were classified by natural properties like fructification, and he was forced to define what he meant by "species" in his work. So far as I am aware, this is the first purely biological definition of "species". It is interesting that in adopting it, Ray ceases to expect that there will be sharp demarcations between species, although he is the first also to assert, as a matter of piety, that species are fixed as they were created, a view later taken for granted by Linnaeus.

> In order that an inventory of plants may be begun and a classification (*divisio*) of them correctly established, we must try to discover criteria of some sort for distinguishing what are called "species". After long and considerable investigation, no surer criterion for determining species has occurred to me than the distinguishing features that perpetuate themselves in propagation from seed. Thus, no matter what variations occur in the individuals or the species, if they spring from the seed of one and the same plant, they are accidental variations and not such as to distinguish a species … Animals likewise that differ specifically preserve their distinct species permanently; one species never springs from the seed of another nor vice versa. [*Historia plantarum generalis,* in the volume published in 1686, Tome I, Libr. I, Chap. XX, page 40 (Quoted in Mayr 1982: 256). The Latin of the definition is *nulla certior occurit quam distincta propagations ex semine.*]

... I would not have my readers expect something perfect or complete; something which would divide all plants so exactly as to include in positions anomalous or peculiar; something which would so define each genus by its own characteristics that no species be left, so to speak, homeless or be found common to many genera. Nature does not permit anything of the sort. Nature, as the saying goes, makes no jumps and passes from extreme to extreme only through a mean. She always produces species intermediate between higher and lower types, species of doubtful classification linking one type with another and having something common with both—as for example the so-called zoophytes between plants and animals. [*Methodus plantarum* of 1682 (quoted in Glass 1959: 35)]

... the number of species being in nature certain and determinate, as is generally acknowledged by philosophers, and might be proved also by divine authority, God having finished his works of creation, that is, consummated the number of species in six days. [Letter (quoted in Greene 1959: 131)]

Nehemiah Grew (1641–1712)

About the same time as Ray, Grew was beginning the observation and description, both by diagram and description, of plant cytology and histology, as we would now call it. He clearly is motivated by inductive generality— what we describe of one plant will be common to all others of its kind. He appeals to the *diairesis* as justification for thinking that definitions are scientific descriptions of real causes and vice versa.

[12. §.] For in looking upon divers *Plants*, though of different *Names* and *Kinds*; yet if some affinity may be found betwixt them, then the *Nature* of any one of them being well known, we have thence ground of conjecture, as to the *Nature* of all the rest. So that as every *Plant* may have somewhat of *Nature individual* to it self; so, as far as it obtaineth any *Visible Communities* with other *Plants*, so far, may it partake of *Common Nature* with those also. [*The Anatomy of Plants* (Grew 1682), *An Idea of a Philosophical History of Plants*, given on January 8, and January 15, 1672 p6]

20. §. From all which, we may come to know, what the *Communities* of *Vegetables* are, as belonging to all; what their *Distinctions*, to such a *Kind*; their *Properties*, to such a *Species*; and their *Peculiarities*, to such Particular ones. And as in *Metaphysical*, or other Contemplative Matters, when we have a distinct knowledge of the *Communities* and *Differences* of Things, we may then be able to give their true *Definitions*: so we may possibly, here attain, to do likewise: not only to know, That every *Plant* Inwardly differs from another, but also wherein; so as not surely to Define by Outward *Figure* than by the Inward *Structure*. What that

is, or those things are, whereby any *Plant*, or Sort of *Plants*, may be distinguished from all others. And having obtained a knowledge of the *Communities* and *Differences* amongst the *Parts* of *Vegetables*; it may conduct us through a *Series* of more facile and probable *Conclusions*, of the ways of their *Causality*, as to the *Communities* and *Differences* of *Vegetation*. ... [p10]

 53. §. The prosecution of what is here proposed, will be requisite, To a fuller and clearer view, of the *Modes* of *Vegetation*, of the *Sensible Natures* of *Vegetables*, and of their more Recluse *Faculties* and *Powers*. First, of the *Modes* of *Vegetation*. For suppose we were speaking of a *Root*; from a due consideration of the *Properties* of any *Organical Part* or *Parts* thereof; 'tis true, that the real and genuine *Causes* may be rendred, of divers and other dependent *Properties*, as spoken generally of the whole *Root*. But it will be asked again, What may be the *Causes* of those *first* and Independent ones? Which, if we will seek, we must do by inquiring also, What are the *Principles* of those *Organical Parts*? For it is necessary, that the *Principles* whereof a Body doth consist, should be, if not all of them the *active*, yet the *capacitating Causes*, or such as are called *Causae sine quibus non*, of its becoming and being, in all respects, both as to *Substance* and *Accidents*, what it is: otherwise, their Existence, in that Body, were altogether superfluous; since it might have been without them: which if so, it might then have been made of any other; there being no necessity of putting any difference, if neither those, whereof it is made, are thought necessary to its Being. ... [p20]

Thomas Burnet (1635–1715)

Burnet defended an old earth, and argued that life may have been formed spontaneously through the heat of the sun on the earth (Gould 1977). His comment that some truths of revelation are undiscoverable runs counter to the natural theology of his day, and it is clear from this passage that he believes species are natural outworkings of "the principles of Life".

 'Tis true, this opinion of the spontaneous Origin of Animals in the first Earth hath lain under some *Odium*, because it was commonly reckon'd to be *Epicurus's* opinion peculiarly; and he extended it not only to all brute Creatures, but to Mankind also, whom he suppos'd to grow out of the Earth in great numbers in several parts and Countries, like other Animals; which is a notion very contrary to the Sacred writings; for they assure us that all Mankind, though diffus'd now through the several parts and Regions of the Earth, rise at first from one Head or single Man and Woman; which is a Conclusion of great importance, and that could not, I think, by the Light of Nature have ever been discover'd. And this makes the *Epicurean* opinion the more improbable, for why should two rise only, if they sprung from the Earth? or how could they rise in their full growth and perfection, as *Adam* and *Eve* did? But as for the opinion of Animals rising out of the Earth at first, that was not at all peculiar to *Epicurus*; The *Stoicks* were of the same mind, and the *Pythagoreans* and the *Ægyptians*, and, I think, all that suppos'd the Earth to rise from a Chaos. Neither do I know any harm in that opinion, if duly limited and stated; for what inconvenience is it, or what diminution of Providence, that there should be the principles of Life as well as the principles of Vegetation in the new Earth? [*The Sacred Theory of the Earth*, Book 2, chapter 1, *The Primeval Earth and Paradise*, pages 182–3 (Burnet 1681 [1684–1689 English edition])]

Joseph Pitton, de Tournefort (1628–1708)

Tournefort appears not to care about "species" versus "variety", so long as they are distinguished by some properties.

> I not only enumerate the several Species of Plants, but often mention what the Botanists call Varieties; not at all solicitous whether they be really the same Species only varied and somewhat diversified, for as they differ in some sensible Qualities, they ought to be distinguished by peculiar Titles. [*Botanical institutions* (*Institutiones rei herbariae*, 1700, earlier published in 1694 as *Elemens botanique* (Tournefort 1716–30: 2)]

Carl Linnaeus (1707–1770, from 1761 Carl von Linné, or Carolus Linnaeus)

There are many myths about Linnaeus that are due to the properties, real or imagined, of the system named after him (Cain 1994; Koerner 1999; Larson 1968; Winsor 2006a). In fact the so-called "Natural System" as it came to be known, was on Linnaeus' own view an artificial one (Cain 1995), and it did not spring forth fully formed from his brow, no matter how much he saw himself as a "second Adam". His view on species was initially fairly standard, based on Ray's definition (perhaps not consciously), and which included fixity. But not, it seems, because of essentialism, but piety.

Linnaeus seems to have defined each species diagnostically, not materially. That is to say, given that species are fixed from the creation, what are the marks of species so we can recognise them? However, he famously discovered what he thought was a novel hybrid species, and so towards the end of his life, began to backpedal the extreme fixity of the earlier writings, calling one such apparent novel species a "daughter of time" (Gustafsson 1979). This was widely known among the botanical community, and was widely accepted. For him, constancy (not essence) of generation was the key to species.

There are as many species as the Infinite Being produced diverse forms in the beginning. [*Species tot sunt diversae quot diversas formas ab initio creavit infinitum Ens, Fundamenta botanica* No. 157, 1736 (quoted in Ramsbottom 1938: 196)]

We reckon as many species as there were diverse forms created in the beginning. [*Species tot numeramus, quot diversae formae in principio sunt creatae, Philosophia Botanica*, 1751 (*loc. cit.*)]

Species are as many as there were diverse [and constant*] forms produced by the Infinite Being; which forms according to the appointed laws of generation, produced more individuals but always like themselves. Therefore there are as many species as there are different forms or structures occurring today. [*Species tot sunt, quot diversas [& constantes*] formas ab initio producit Infinitum Ens; quae formae, secundum generationis inditas leges, produxere plures, at sibi semper similes. Ergo species tot sunt, quot diversae formae s. structurea hodienum occurrant. Classes Plantarum*, 1738 (*loc.cit.*). *Added in 1764, see *Genera Plantarum* I: ¶5]

The principle being accepted that all species of one genus have arisen from one mother through different fathers, it must be assumed:

1) That in the beginning the Creator created each natural order only with one plant with reproductive power.
2) That by their various mixings different plants have arisen which belong to the mother's natural order as they are similar to the mother with regard to their fructifications, and are, as it were, species of the order, i.e., genera.
3) We may assume that plants have arisen within the orders, i.e. by genera of one order, may mix with each other. In this way there will arise species that should be referred to the mother's genus as her daughters. [*Pralectiones* (Lectures, 1744), quoted in Larson, (1967: 317)]

We say there are as many genera as there are similarly constituted fructifications of different natural species. [*Genera tot dicimus, quot similes contructae fructifications proferunt diversae Species naturales. Fundamenta Botanica* 1736, No 159 (quoted in Ramsbottom 1938: 197)]

Every genus is natural, created as such in the beginning, hence not to be rashly split up or stuck together by whim or according to anyone's theory. [*Genus omne est naturale, in primordio tale creatum, hinc pro libitu & secundem cujuscimque theoriam non proterve discindendum aut conglutinandum. Systema naturae*, 1735, (quoted in Ramsbottom 1938: 197)]

Species are most constant, since their generation is a true continuation. [*Species constantissimae sunt, cum earum generatio est vera continuatio. Systema naturae,* 1735 (quoted in Ramsbottom 1938: 197)]

There are as many varieties as there are different plants, produced from the seed of the same species. [*Varietates tot sunt, quot differentes plantae ex ejusdem speciei semine sunt productae. Philosophia Botanica* 1751 (quoted in Ramsbottom 1938: 199)]

Is the plant [*Thalictrum lucidum*] sufficiently distinct from *T. flavum?* It seems to me a daughter of time. [*Planta, an satis distincta, a* T. flavo? *Videtur temporis filia. Species plantarum* 1753 (quoted in Ramsbottom 1938: 201)]

Pierre-Louis Moreau de Maupertuis (1698–1759)

The first edition of the *Systema Naturae* was published in 1735. Maupertuis published the first scientific account of novel species appearing via inherited changes in 1743, a mere eight years later. It is almost as if it took the broad publication of fixism for transformationism to be a formal possibility. Alas, Maupertuis' mechanisms were crude, and he was denounced by Diderot and Voltaire, and failed to make much impact on his contemporaries. Still, Maupertuis noted that heredity was (i) particulate, and (ii) bisexual equally (Gasking 1967). Moreover, he also noted something like a Mendelian assortment.

… it is not rare to find dogs with a fifth digit on the hind feet, although it is usually detached from the bone and without articulation. Is this fifth digit of the hindfeet an extra one, or is it nothing other than a digit lost from variety to variety in the whole species, and which tends to reappear from time to time?

Could we not explain in this manner [of fortuitous changes] how the multiplication of the most dissimilar species could have sprung from just two individuals? They would owe their origin to some fortuitous productions in which the elementary parts [of heredity] deviated from the order maintained in the parents. Each degree of error would have created a new species, and as a result of repeated deviations the infinite diversity of animals that we see today would have come about. [*Systéme de la Nature* 2:164 (quoted in Terrall 2002: 338)]

… all the matter we now see on the surface of our Earth was once fluid, whether dissolved in water or melted by fire. Now, in this fluid state, the matter of our globe was in the same situation as the liquors in which the elements that produce animals swim: and metals, minerals, precious stones were much easier to form than the least organicized insect. The least active particles of matter will have formed metals and stones [*marbres*]; the most active formed animals and man.

The only difference between these productions is that some continue in a fluid state, and in others the hardening of the matter containing their elements does not permit new productions. [*Ibid* 2:169 (quoted in Terrall 2002: 339)]

After such a flood or fire, new unions of elements, new animals, new plants or rather entirely new things could reproduce themselves. [*Ibid* 2:170 (*loc. cit.*)]

Georges-Louis Leclerc, Comte de Buffon (1707–1788)

Buffon is very influential via his multi-volume *Histoire Naturelle*, which was owned or read by every gentleman and naturalist of his generation, and was still influential a century later. He had two distinct periods or views of species—one in which they were Lockean conveniences, and one in which they were defined by mutual sterility (Sloan 1979, 1985; Farber 1971; Lovejoy 1959). His "biological" definition is not, as Mayr and others said, purely about interfertility, but also about the generation of progeny that are similar to their parents (implied in the use of the term *invariable* in the second excerpt). Moreover, his "species" is more akin to the Linnaean genus or higher, so reproductive interfertility is much broader on his view than the modern biospecies conception. He believed that Linnaean species evolved by degeneration and the action of habitat and soil from a "primary stock" (*premiere souche*), and so, for example, all great cats were a single species that could be backbred to the original stock.

The error consists in a failure to understand nature's processes (*marche*), which always take place by gradations (*nuances*). … It is possible to descend by almost insensible degrees from the most perfect creature to the most formless matter. … These imperceptible shadings are the great work of nature; they are to be found not only in the sizes and forms, but also in the movements, the generations and the successions of every species. … [Thus] nature, proceeding by unknown gradations, cannot wholly lend herself to these divisions [into genera and species]. … There will be found a great number of intermediate species, and of objects belonging half in one class and half in another. Objects of this sort, to which it is impossible to assign a place, necessarily render vain the attempt at a universal system. …

In general, the more one increases the number of one's divisions, in the case of the products of nature, the nearer one comes to the truth; since in reality individuals alone exist in nature. [*Histoire naturelle*, (1749: 12, 13, 20, 38; quoted in Lovejoy 1936: 230)]

We should regard two animals as belong to the same species if, by means of copulation, they can perpetuate themselves and the likeness of the species; and we should regard them as belonging to different species if they are incapable of producing progeny by the same means. Thus the fox will be known to be a different species from the dog if it proves to be a fact that from the mating of a male and female of these two kinds of animals no offspring is born; and even if there should result a hybrid offspring, a sort of mule, this would suffice to prove that fox and dog are not of the same species—inasmuch as this mule would be sterile (*ne produirait rien*). For we have assumed that, in order that a species might be constituted, there was necessary a continuous, perpetual and unvarying reproduction (*une production continue, perpétuelle, invariable*)—similar, in a word, to that of other animals. [*Histoire naturelle* Vol. 2 (1749), 10 (Lovejoy 1959: 93f)]

These physical genera [of sheep, oxen and dogs] are, in reality, composed of all the species, which, by our management, have been so greatly variegated and changed; as have all those species, so differently modified by the hand of man, have but one common origin in Nature, the whole genus ought to constitute but a single species. ["The Sheep", *Histoire Naturelle*, vol 6, (1756), 221 (quoted in Greene 1959: 147)]

A sparrow or warbler has perhaps twenty times as many relatives as an ostrich or a turkey; for by the number of relatives I understand the number of related species that are sufficiently alike among themselves to be considered side branches of the same stem, or at least ramifications of stems that grow so closely together that one can suspect that they have a common root, and can assume that originally they all sprang from this root, of which one is reminded by the large number of their shared similarities; and these related species probably have separated only through the influence of climate, food, and the procession of years, which brings into being every realizable combination and allows every possibility of variation, perfection, alteration, and degeneration to become manifest. [*Histoire naturelle des oiseaux* (1770) (Stresemann 1975: 56 translating p75 in the original.)]

John Hunter (1728–1793)

Hunter was a widely respected doctor and naturalist in his day. His views expressed here are evidence of the advanced technical views of the British scientific establishment at the latter end of the 18th century regarding species. He appears to be the first author to claim that wolves, dogs and other canines might be the same species, although he included the jackal as well, which modern biologists do not.

The true distinction between different species of animals must ultimately, as appears to me, be gathered from their incapacity of propagating with each other an offspring capable again of continuing itself by frequent propagations: thus the

Horse and Ass beget a Mule capable of copulation, but incapable of begetting or producing offspring. If it be true, that the mule has been known to breed, which must be allowed to be an extraordinary fact, it will by no means be sufficient to determine the Horse and Ass to be of the same species; indeed, from the copulation of Mules being very frequent, and the circumstance of their breeding very rare, I should rather attribute it to a degree of monstrosity in the organs of the Mule which conceived, not being those of a mixed animal, but those of the Mare or female Ass. This is not so far-fetched an idea, when we consider that some true species produce monsters, which are a mixture of both sexes, and that many animals of distinct sex and incapable of breeding at all.

If then we find nature in its greatest perfection deviating from general principles, why may not it happen likewise in the production of Mules, so that sometimes a Mule shall breed from the circumstances of its being a monster respecting mules? [Hunter, John. 1787. Observations Tending to Shew That the Wolf, Jackal, and Dog, are All of the Same Species. *Philosophical Transactions of the Royal Society of London* (Hunter 1787: 253)]

Immanuel Kant (1724 – 1804)

Kant famously said of the living world that it inevitably seemed to be driven by final causes and thus there would never be a Newton "who shall make comprehensible by us the production of a blade of grass according to natural laws which no design has ordered" (*Critique of Judgment,* §75). However, he was particularly perspicacious when it came to classification in biology, although he tended to accept the racial classifications formalised by Blumenbach. He believed that what maintained species integrity was the "seed" (*Keime*) that exists *in potentia* in species, but that classification needed to be genealogical rather than based on definitional properties (Richards 2000).

> In the animal kingdom the natural classification into species and variety is grounded in the common law of propagation, and the unity of the species is nothing other than the unity of the generative force, which holds thoroughly for a certain manifold of animals. Accordingly, Buffon's rule that animals which produce fertile offspring with one another, (regardless of the difference in form there may be between them) belong to one and the same physical species, is actually viewed merely as the definition of a natural species of animal in general, in contrast to all scholastic species of animals. The scholastic classification is made according to classes and orders animals according to similarities. The natural classification, however, is based on lines of descent and orders them according to relationships with respect to generation. The former accomplishes a scholastic

system for the memory, the latter a natural system for the understanding; the former intends only to bring the creatures under titles, the latter to bring them under laws. [*On the Different Races of Man*, 1775[2]]

For animals whose variety is so great that an equal number of separate creations would have been necessary for their existence could indeed belong to a *nominal family grouping* [*Nominalgattung, lit.* nominal species] but never to a *real one*, other than one as to which at least the possibility of descent from a single common pair is to be assumed. ... [Otherwise] the singular compatibility of the generative forces of **two** species (which, although quite foreign as to origins, yet can be fruitfully mated with each other) would have to be assumed with no other explanation than that nature so pleases. If, in order to demonstrate this latter supposition, one points to animals in which crossing can happen despite the [supposed] difference of their original stems, he will in every case reject the hypothesis and, so much the more because such a fruitful union occurs, infer the unity of the group, as from the crossing of dogs and foxes, etc. The *unfailing inheritance* of peculiarities of both parents is thus the only true and at the same time adequate touchstone of the unity of the group from which they have sprung: namely the original seeds [*Keime*] inherent in this group developing in a succession of generations without which those hereditary variations would not have originated and would presumably not *necessarily* have *become* hereditary. [Determination of the concept of human races" (1785) (trans. in Greene 1959: 233)]

That the manifold respects in which individual things differ do not exclude identity of species, that the various species must be regarded merely as different determinations of a few genera, and these, in turn, of still higher genera, and so on; in short that we must seek for a certain systematic unity of all possible empirical concepts, in so far as they can be deduced from higher and more general concepts—this is a logical principle, a rule of the Schools, without which

---

2 Im Thierreiche gründet sich die Natureintheilung in Gattungen und Arten auf das gemeinschaftliche Gesetz der Fortpflanzung, und die Einheit der Gattungen ist nichts anders, als die Einheit der zeugenden Kraft, welche für eine gewisse Mannigfaltigkeit von Thieren durchgängig geltend ist. Daher muß die Büffonsche Regel, daß Thiere, die mit einander fruchtbare Jungen erzeugen, (von welcher Verschiedenheit der Gestalt sie auch sein mögen) doch zu einer und derselben physischen Gattung gehören, eigentlich nur als die Definition einer Naturgattung der Thiere überhaupt zum Unterschiede von allen Schulgattungen derselben angesehen werden. Die Schuleintheilung geht auf Klassen, welche nach Ähnlichkeiten, die Natureintheilung aber auf Stämme, welche die Thiere nach Verwandtschaften in Ansehung der Erzeugung eintheilt. Jene verschafft ein Schulsystem für das Gedächtniß; diese ein Natursystem für den Verstand: die erstere hat nur zur Absicht, die Geschöpfe unter Titel, die zweite, sie unter Gesetze zu bringen. [*Gesammelte Werke* II: 429. Translation by Mark Fisher, used with permission. A version of this can be found also in (Dobzhansky 1962: 93, from which I found this passage).]

there can be no employment of reason. [*Critique of Pure Reason* B679f (Second edition, 1787) (Kant 1933)]

The logical principle of genera, which postulates identity, is balanced by another principle, namely that of *species*, which calls for manifoldness and diversity in things, notwithstanding their agreement as coming under the same genus, and which prescribes to the understanding that it attend to the diversity no less than to the identity. [B682, *op. cit.*]

For if there were no *lower* concepts, there could not be *higher* concepts. Now the understanding can have knowledge only through concepts, and therefore, however far it carries the process of division, never through mere intuition, but always again through *lower* concepts. The knowledge of appearances in their complete determination, which is possible only through the understanding, demands an endless progress in the specification of our concepts, and an advance to yet other remaining differences, from which we have made an abstraction in the concept of the species, and still more so in that of the genus. [B684, *op. cit.*]

... all differences of species border upon one another, admitting of no transition from one to another *per saltum,* but only through all the smaller degrees of difference between them." [B687, *op. cit.*]

For in the first place, the species in nature are actually divided, and must therefore constitute a *quantum discretum.* ... And further, in the second place, we could not make any determinate empirical use of this law, since it instructs us only in quite general terms that we are to seek for grades of affinity, and yields no criterion whatsoever as to how far, and in what manner, we are to prosecute the search for them" [B689, *op. cit.*]

In order to see that a thing is only possible as a purpose, that is to be forced to seek the causality of its origin, not in the mechanism of nature, but in a cause whose faculty of action is determined through concepts, it is requisite that its form be not possible according to mere natural laws... The *contingency* of its form in all empirical natural laws in reference to reason affords a ground for regarding its causality as possible only through reason. [*Critique of Judgement* in 1790 §64 (second edition in 1793: Kant 1951)]

Hence it is only so far as matter is organized that it necessarily carries with it the concept of a natural purpose, because this its specific form is at the same time a product of nature. ...

If we have once discovered in nature a faculty of bringing forth products that can only be thought by us in accordance with the concept of final causes, we go further still. We venture to judge that things belong to a system of purposes which yet do not (either in themselves or in their purposive relations) necessitate our seeking for any principle of their possibility beyond the mechanisms of causes working blindly. For the first idea, as concerns its ground, already brings us beyond the world of sense, since the unity of this supersensible principle must be regarded as valid in this way, not merely for certain species of natural beings, but for the whole of nature as a system. [§67, *op. cit.*]

Species and genus [of logic—*JSW*] are not distinguished in natural history (which has only to do with ancestry and origin). Only in the description of nature, since it is a matter of comparing distinguishing marks, does this distinction come into play. What is species here must there often be called only race. [1785, *Gesammelte Schriften*, VIII, 100n (quoted in Greene 1959: 372n)]

Many animal species resemble one another according to a certain common scheme, which scheme seems to lie at the foundation not only of the structure of their bones but also of the ordering of their other parts, so that the proliferation of species might arise according to a simple outline: the shortening of one part or the lengthening of another, the development of one part or the atrophy of another. This possibility produces a faint ray of hope that something might be done with the principle of mechanism, without which no natural science can generally be constituted. This analogy of forms—insofar as they seem to have been produced, despite their differences, according to common archetypes [*Urbilde*]—strengthens the suspicion of a real relationship of these forms by reason of their birth from a common, aboriginal mother [*Urmutter*]. [*Kritik der Urteilskraft* (Werke, vol. 5, p. 538 (A363–64, B368–69), quoted in Richards 2000: 28)]

Michel Adanson (1727–1806)

Adanson's views on species have been widely misrepresented (Winsor 2004) to require that all (total) evidence is needed to define and identify a species. In fact he did not say this, but held that the more evidence one has the more reliable the diagnosis. He clearly thought that constancy of character (which may or may not be the same as fixism) was the defining property of species.

It was necessary to seek in nature for nature's system, if there really was one. With this aim, I examined plants in all their parts, without omitting one, from roots to embryo, folding of leaves in the bud, manner of sheathing, development, position, and folding of the embryo and radicle in the seed relative to the fruit; in a word, a number of features to which few botanists pay attention. [*Family of plants* (*Familles des Plantes*, 1763, 1764) (Morton 1981: 303)]

What is sufficient to constitute the genera of certain families is not sufficient for other families, and neither the same parts nor the same number of these parts invariably furnish these [constituent] parts in each family. [*Op. cit.*, (Morton 1981: 305)]

Although it is very difficult, not to say impossible, to give an absolute and general definition of any object of natural history whatever, one could say rather exactly that there are as many species as there are different individuals among them, different in any (one or more) respect, constant or not, provided they are definitely perceptible and taken from parts or qualities where those differences appear to be most naturally placed in accordance with the particular character of each family.[3] [*Op. cit.*, (translation from Stafleu 1963: 185)]

The moderns define a species of plant as a collection of several individuals which resemble each other perfectly, yet not in everything, but in the essential parts and qualities, without, however, giving attention to the differences caused in these individuals either by sex or accidental varieties.

According to Linnaeus (Phil. bot., p. 99) "the species of plants are natural and constant, as their propagation either by seeds or cuttings is only a continuation of the same species. Individuals die, but the species does not."

But we wish to make a distinction between reproduction by seed and that by shoots, offsets, corms, cuttings, suckers or by grafting. These last simply continue the individual from which they are taken and consequently are opposed to the production of new species in plants; *whereas seeds are the source of a prodigious number of varieties, sometimes so changed that they may pass for new species.* He cites, among other examples:

"In 1715 Marchant found in his garden a new species of Mercurialis and the following year it came from self-sown seed; again, four resembled the parent and two were so different that he made another species of Mercurialis. These two new plants were cultivated and continued to grow each year."

It is well known that without foreign fecundation in plants that reproduce by seed, similar changes are induced either by reciprocal fecundation of two different individuals or owing to cultivation, the soil, the climate, dryness or moisture, light or shade, etc. These changes are more or less prompt, more or less durable, disappearing in one generation or perpetuating themselves through several generations, according to the number, the force, the duration of the causes which united to form them, etc., according to the nature, the disposition, the customs,

---

3　Definition de l'Espèce.

Ainsi, quoiqu il soit très-dificil, pour ne pas dire impossible, de donor une définition absolue & générale d'aucun objet de l'Hist. nat. on pouroit dire assez exactemant qu'il existe autaunt d'Espèces, qu'il i a d'Individus diférans entreaux, d'une ou de plusieurs diférances quelkonkes, constantes ou non, pourvo qu'eles soient très-sensibles, & ti-rées des parties ou qualités où ces diférances paroissent plus naturelemant placées, selon le génie ou les moeurs propres à chaque Famille; ... (Adanson 1763: clxviij)

so to speak, of each plant, for it is to be noted that some families do not vary except in the roots, others in the leaves, others in height, pubescence, and color, whereas others change more easily their flowers or their fruit.

It is difficult to define a primitive species and which those are which have originated by successive reproduction or been changed by accidental causes. It is without doubt for this reason that we do not find nowadays a number of plants described by ancient botanists; they have disappeared, either by returning to primitive forms or by changing their form in the multiplication of species. For this reason the ancients knew fewer species; time has brought novelties! And for the same reason future botanists will be *overwhelmed* by the number of species and be obliged to abandon them and be *reduced solely to genera!* [*Familles des Plantes*, 1763 (quoted in Britton 1908: 225f)]

Species: collection of all objects which nature separates individually from each other as so many isolated entities existing separately and which the imagination or the free and creative opinion of man unites *idéalement* each time that he finds an almost complete resemblance or a resemblance at any rate greater than with any other group, a collection to which he gives the name species. [Handwritten definition in the fifth volume of his copy of the *Encyclopédie* of Diderot (translated in Stafleu 1963: 186)]

Antoine-Laurent de Jussieu (1748–1836)

Jussieu is one of the more significant French taxonomists of the period (Stevens 1994). His view of species is very much the model of the Generative Conception. However, he has a strong typological view, ameliorated by his acceptance of the *scala naturae* view of plenitude, or the completeness of the existence of all possible kinds.

Just so many plants agreeing in all their parts, or being consistent in their universal character, and born from and giving birth to those of like nature, are the individuals together constituting a species, [a term] wrongly used in the past, now more correctly defined as the perennial succession of like individuals, successively reborn by continued generation. [*Genera plantarum secundum ordines naturalis disposita* (1789), (translations in Stevens 1994: 292)]

… the species must first be known, and defined by its proper signs: [it is] a collection ["adhesio"] of beings that are alike in the highest degree, never to be divided, but simple by unanimous consent [and] simple by the first and clearest law of Nature, which decrees that *in one species are to be assembled all vegetative beings or individuals that are alike in the highest degree in all their parts, and that are always similar ["conformia"] over a continued series of generations,* so that any individual whatever is the true image of the whole species, past, present, and future. ["The sure knowledge of species" (*op. cit.*, 356f), Italics original] [4]

[The natural method] ... links all kinds of plants by an unbroken bond, and proceeds step by step from simple to composite, from the smallest to the largest in a continuous series, as a chain whose links represent so many species or groups of species, or like a geographical map on which species, like districts, are distributed by territories and provinces and kingdoms. [*Op. cit.*, 355]

Charles Bonnet (1720–1793)

Bonnet also held a Great Chain of Being view of life, indeed he was probably the culmination of that view (Anderson 1976; Lovejoy 1946). Immediately before Lamarck, he represents the last atemporal version of the *scala naturae*.

If there are no cleavages in nature, it is evident that our classifications are not hers. Those which we form are purely nominal, and we should regard them as means relative to our needs and to the limitations of our knowledge. Intelligences higher than ours perhaps recognize between two individuals which we place in the same species more varieties than we discover between two individuals of widely separated genera. Thus these intelligences see in the scale of our world as many steps as there are individuals. [*Contemplation de la Nature*, 2nd edn, 1769, I, p28, (quoted in Lovejoy 1936: 231)]

Jean Baptiste de Lamarck (1744 –1829)

Lamarck, early in his career, wrote a perfectly orthodox definition of "species" for an encyclopedia, but the real novelty of his views arises when he takes the early Buffon seriously and denies that species exist, because if they existed, they would be constant (Hull 1984b; Packard 1901; Jordanova 1984; Barthélemy-Madaule 1982; Burkhardt 1985).

Species: in botany as in zoology, a species is necessarily constituted of the aggregation of similar individuals which perpetuate themselves, the same, by reproduction. I understand similarity in the essential qualities of the species, because the individuals which constitute it offer frequently accidental differences which give rise to varieties and sometimes sexual differences, which belong however to the same species, as the male and female hemp, in which all the individuals constitute the cultivated hemp. Thus, without the constant

---

4 The Latin of the definition is (Jussieu 1964: xxxvij):
  ... *in unam speciem colligenda sunt vegetantia seu individua omnibus suis partibus simillima & continuatâ generationum serie semper conformia...*

reproduction of similar individuals, there could not exist a true species. [*Encyclopedie Methodique*, Vol. 2, 1786 (quoted in Britton 1908: 227)]

It is not a futile purpose to decide definitely what we mean by the so-called *species* among living bodies, and to enquire if it is true that species are of absolutely constancy, as old as nature, and have all existed from the beginning just as we see them to-day; or if as a result of changes in their environment, albeit extremely slow, they have not in the course of time changed their characters and shape.

...

Let us first see what is meant by the name of species.

Any collection of like individuals which were produced by others similar to themselves is called a species.

This definition is exact: for every individual possessing life always resembles very closely those from which it sprang; but to this definition is added the allegation that the individuals composing a species never vary in their specific characters, and consequently that species have an absolute constancy in nature.

It is just this allegation that I propose to attack, since clear proofs drawn from observation show that it is ill-founded.

[*Zoological philosophy* (Lamarck 1809; English translation Lamarck 1914: 35)]

Thus, among living bodies, nature, as I have already said, definitely contains nothing but individuals which succeed one another by reproduction and spring from one another; but the species among them have only a relative constancy and are only invariable temporarily. [p44]

Nevertheless, to facilitate the study and knowledge of so many different bodies it is useful to give the name species to any collection of like individuals perpetuated by reproduction without change, so long as their environment does not alter enough to cause variations in their habits, character and shape. [*loc. cit.*]

Christian Leopold Freiherr von Buch (1774–1853)

A geologist and paleontologist, von Buch visited the Canary Islands in 1815, and in the course of describing the flora, made the following comment, which was quoted by Hans Gadow in a volume published for the fiftieth anniversary of the *Origin*, and which has been widely cited. Mayr, in particular, has claimed von Buch as a precursor to the allopatric theory of speciation. It is not clear that this was in fact his actual view—he appears to think that conditions cause changes but interbreeding would cause a regression to the mean, as Trémaux later did (see below).

Upon the continents the individuals of the genera by spreading far, form, through differences of the locality, food and soil, varieties which finally become constant as new species, since owing to the distances they could never be crossed with other varieties and thus be brought back to the main type. Next they may again, perhaps upon different roads, return to the old home where they find the old type likewise changed, both having become so different that they can interbreed no longer. Not so upon islands, where the individuals shut up in narrow valleys or within narrow districts, can always meet one another and thereby destroy every new attempt towards the fixing of a new variety. [*Physikalische Beschriebung der canarischen Inseln* 1825, (von Buch 1877: 345; translated and quoted by Gadow 1909: 326n)]

Isidore Geoffroy Saint-Hilaire (1805–1861)

Geoffroy was also a transformationist, influenced by but not following Lamarck (Appel 1987). He offers a relative stability criterion instead of Lamarck's species denial—they are real, but not permanent.

The species is a collection or a succession of individuals characterized by a whole of distinctive features whose transmission is natural, regular and indefinite in the current order of things. [*Histoire naturelle génerale des règnes organiques*, 1859 (vol. 2: 437, quoted in Lherminer and Solignac 2000: 156)[5]]

Georges Leopold Chrétien Frédéric Dagobert, Baron Cuvier (1769–1832)

Cuvier's view is formulated in part in direct opposition to Lamarck's, and is clearly in the Generative Conception tradition (Eigen 1997; Coleman 1964). However, Cuvier, while he did not think species either appeared gradually or were immune from extinction, was a species fixist while they existed. Cuvier's definition was widely quoted by biologists throughout the 19th century as the authoritative definition. He was a famous opponent of speculative hypotheses and always claimed to hold fast to evidence and not speculate. Hence he refused to hypothesise about the origin of new species in each faunal succession.

My research assumes the definition of species which serves as the basic use made of the term, understanding that the word species means *the individuals who descend from one another or from common parents and those who resemble them as much as they resemble each other.* Thus, we call varieties of a species only

---

5 L'espèce est une collection ou une suite d'individus caractérisés par un ensemble de traits distinctifs dont la transmission est naturelle, régulière et indéfinie dans l'ordre actuel des choses.

those races more or less different which can arise from it by reproduction. Our observations on the differences among the ancestors and the descendants are therefore for us the only reasonable rule, because all others would take us back to hypotheses without proofs.

Now, by taking the word *variety* in this way, we observe that the differences which constitute it depend on fixed circumstances and that their extent increases according to the intensity of these circumstances. [*Règne Animal*, i, 19 (Cuvier 1812)[6]]

I do not pretend that a new creation was required for calling our present races of animals into existence. I only urge that they did not anciently occupy the same places, and that they must have come from some other part of the globe. [*Essay*, 1818 English translation, 128 (quoted in Greene 1959: 363n)]

James Cowles Prichard (1786–1848)

Prichard was an influential anthropologist who argued strongly for monogenism (that all humans were a single species with a single origin). He, however, holds that variety is constrained by natural limits marked by constant characters.

[Linnaeus' classes are arbitrary and artificial.] Not so in the case of species. Here the distinction is formed by nature, and the definition must be constant and uniform, or it is of no sort of value. It must coincide with Nature.

Providence has distributed the animated world into a number of distinct species, and has ordained that each shall multiply according to its kind, and propagate the stock to perpetuity, none of them ever transgressing their own limits, or approximating in any great degree to others, or ever in any case passing into each other. Such confusion is contrary to the established order of Nature.

The principle therefore of the distinction of species is constant and perpetual difference. [*Researches into the Physical History of Man* (pp7–8, quoted in Greene 1959: 239)]

William Whewell (1794–1866)

Whewell was a friend of Darwin's, and also of Hooker's and Lyell's, and his philosophy was something Darwin took into account. However, he was famously a fixist, although not an essentialist, allowing that types do not have

---

6   Under a section entitled "Lost species are not varieties of living species":
     Cette recherche suppose la définition de l'espèce qui sert de base à l'usage que l'on fait de ce mot, savoir que l'espèce comprend *les individus qui descendent les uns des autres ou de parens communs, et ceux qui leur ressemblent autant qu'ils se ressemblent entre eux.* [From <http://www.mala.bc.ca/~johnstoi/cuvier.htm>, accessed 20 April 2004]

fixed definitions. Instead, they are typical models that most examples of the group have the same differentiae as. His notion that members of a type could be typical rather than total was attacked by Mill (see below). He was not only a species realist, but probably *the* species realist.

> ... *species have a real existence in nature*, and a transition from one to the other does not exist. (Whewell 1837, v. 3, 626; quoted in Hull 1973: 68)

> Natural groups are given by Type, not by Definition. And this consideration accounts for that indefiniteness and indecision which we frequently find in the descriptions of such groups, and which must appear so strange and inconsistent to anyone who does not suppose these descriptions to assume any deeper ground of connection than an arbitrary choice of the botanist. Thus in the family of the rose-tree, we are told that the *ovules* are *very rarely* erect, the *stigmata usually* simple. Of what use, it might be asked, can such loose accounts be? To which the answer is, that they are not inserted in order to distinguish the species, but in order to describe the family, and the total relations of the ovules and the stigmata of the family are better known by this general statement ... .

> The type-species of every genus, the type-genus of every family, is then one which possesses all the characters and properties of the genus in a marked and prominent manner. The type of the Rose family has alternate stipulate leaves, wants the albumen, has the ovules not erect, has the stigmata simple, and besides these features, which distinguish it from the exceptions or varieties of its class, it has the features which make it prominent in its class. It is one of those which possess clearly several leading attributes; and thus, though we cannot say of any one genus that it *must* be the type of the family, or of any one species that it *must* be the type of the genus, we are still not wholly to seek; the type must be connected by many affinities with most of the others of its group; it must be near the center of the crowd, and not one of the stragglers. [(Whewell 1858)]

Johann Wolfgang Goethe (1749–1832)

Goethe's and the so-called "ideal morphologists'" primary interests were with the relation between form and function (Lenoir 1987; Opitz 2004). They rarely make comment on the nature of taxa, and for instance, Goethe's *Metamorphosis of Plants* (1790) (Goethe, Tobler, and Arber 1946), is not about metamorphosis of species to species but leaf into flower and similar mathematical formal transformations between parts of plants. The Type here, which is later itself transformed into Owen's "Archetype", is not a historical entity at all (Amundson 2005).

We share the author's [d'Alton's] conviction that there is a general type. … And we also believe in the external plasticity of all phenomenal forms. In this context it is, however, relevant to note that, once certain forms have come to exist as definite genera, once they have taken on a specific individuality, they persist with obstinacy over many generations, and remain essentially true to themselves even in their greatest deviations. [Review of Eduard d'Alton (1772–1840), and Christian Heinrich Pander (1794–1865), *Vergleichende Osteologie der Säugetiere* (*Comparative osteology of mammals*, 1821–1831) (Translated in Wells 1978: 34)]

Archbishop Richard Whately (1787–1863)

Whately's *Elements of Logic*, first published in 1826, when Darwin was still an undergraduate, revitalised the teaching of the diairetic logic in England and the rest of the English-speaking world. It went through at least a dozen editions, and was still being used at the turn of the twentieth century. As such it might be expected to represent the standard view about the relation between formal logic and natural classifications just before Darwin voyaged. If essentialism will be in evidence anywhere, it must be here. It is not.

Whately's book inspired both a response from George Bentham, Jeremy's nephew and later one of the greatest of British botanists, and John Stuart Mill's *A System of Logic* (1843), as well as being the standard text and authority (or target) in logical matters well into the 1890s. In every edition from first to fifth, the following passage remains unchanged.

> … if anyone utters such a proposition as … "Argus was a mastiff," to what head of Predicables would such a Predicate be referred? Surely our logical principles would lead us to answer, that it is the *Species*; since it could hardly be called an Accident, and is manifestly no other Predicable. And yet every Naturalist would at once pronounce that Mastiff, is no distinct Species, but is only a *variety* of the Species Dog. …
>
> … the solution of the difficulty is to be found in the peculiar technical sense … of the word "Species" when applied to *organized Beings*: in which case it is always applied (when we are speaking strictly, as naturalists) to individuals as are supposed to be *descended from a common stock*, or which *might* have so descended; *viz.* which resemble one another (to use M. Cuvier's expression) as much as those of the same stock do. [*Elements of Logic* (1826) Bk IV, ch. 5 §1 (Whately 1875: 183, fifth edition used)]
>
> [The fact of two organisms being the same species] being one which can seldom be *directly* known, the consequence is, that the *marks* by which any Species of

Animal or Plant is *known*, are not the very *Differentia* which *constitutes* that Species. [*Op. cit.*, p184f]

George Bentham (1800–1884)

The nephew of the great utilitarian philosopher, Jeremy Bentham, George Bentham, critiqued Whately in detail (Bentham 1827), and introduced the first logical tree diagrams applied to botanical classifications in that book. He thereafter became a botanist, and a good one. Although he didn't theorise much *qua* botanist, it seems, he made some comments on our topic. He applies the genealogical criterion, or something potentially like it. Note that in the later passage, he does not, as McOuat (2003) claims, proffer an essentialist account of the history of species conceptions so much as a standard fixist one. That there are limits to variation does not mean there is an essence doing the constraining. I believe Bentham is not concerned with essences here, and in the quotations taken from the *Outline* by McOuat (his note 24), where Bentham talks about definitions, I think he is referring to identification rather than material essentialism (see sections 4 and 8 below).

> 12. Division (logical)—the converse of generalization. From a general class, constituted by the combination of any number of properties, the forming subclasses, by the additional consideration of other properties applicable only to a part of the individuals to which the general class in question refers. Thus, the class *animal* having been founded on the consideration of the property of *animation*, by the additional consideration of the property of *rationality*, we form the two subclasses of *man*—consisting of animals which are possessed of that property, and of *brutes*—or those which are deprived of it. ... [Chapter IV, *Outline of a New System of Logic* (Bentham 1827: 58)]
>
> 13. Analysis—the converse of synthesis. It operates on individual ideas, in the same manner that logical division does on classes. Man, for ex., may be divided *logically* into Negroes and Whites. A man may be *analysed* (mentally as well as physically) into his *head, neck, body,* and *limbs.* [*Ibid*, p59]
>
> As to predicating the "whole essence" of a subject, in one term, that is impossible, unless that term be a strict synonym. Dr. Whately does not appear to have been aware of this; he implies that if we predicate the genus, we predicate a part of the essence; if the species, we predicate the whole essence; considering *species* in the sense in which naturalists employ the word, in which case it is, in fact, a logical genus with reference to individuals. [*Ibid*, p69]
>
> 4. Definition—the exhibition of the genus to which a collective idea belongs, and of the properties which distinguish it from all other species of the same

genus. That is the mode which Aristotelians term "Definitio per genus et differentiam," and which Dr. Whately has denominated "Essential Definition." It is applicable only to common names, and, when practicable, is the most exact of all modes of expression. Ex. *Homo est animal rationale.* The *characteristic phrases* of naturalists [i.e., the short species definitions in Linnaean taxonomy—*JSW*] are all definitions. [Chapter VI, p 79, *Ibid*]

8. Description is a detailed exposition of *accidental* properties; that is, of those of which the exhibition is not necessary for the distinguishing of the object from all others which are not designated by the same name. [*Ibid*, p82]

A botanist visits a country with whose productions he is as yet unacquainted; he sees a number of plants which resemble one another very strongly, and which differ considerably from any other plants which he has seen or heard of; he discovers successively several of these *sets* of plants, and by *generalization* he forms as many new *species*, characterized by the properties he has *observed* in these several individual plants. [*Ibid*, Chapter VII p102–3]

... the agriculturist may divide Live Stock according to the *usefulness for cultivation* of the different animals, as compared with the *expense of keep, prime cost*, &c. The naturalist divides animals according to their *conformation*; the commissary may derive his divisions of animals from their usefulness as *furnishing food*, or as *beasts of draught*, &c. [*Ibid*, p106]

... all the great masters of science [agree that a species is] ... a collection of individuals which, by their resemblance to each other, or by other circumstances, we are induced to believe are descended or *may have* descended from one individual or a pair of individuals. ... In the practical application of such a definition to individual cases, in the determination of the specific identity or distinctness of two given plants, there is, indeed, among botanists the most deplorable diversity of opinion. [Anonymous review of de Candolle's *Geographie Botanique Raisonée* in 1856 (quoted in Stevens 1997: 360)]

The great thunderbolt [the Darwinian theory] had, indeed, been launched, but had not yet produced its full effect. We systematists, bred up in the doctrine of the fixed immutability of species within positive limits, who had always thought it one great object to ascertain what those limits were and by what means species, in their neverending variation and constant attempts to overstep those limits, were invariably checked and thrown back in their domain, we might have felt disposed to resist the revolutionary tendency of the new doctrines, but we felt shaken and puzzled. [Presidential Address to the Linnean Society, 1871 (quoted in McOuat 2003: 204f, corrected)]

John Stuart Mill (1806–1873)

Mill's *A System of Logic* needs no introduction. It is still being discussed today in logic and the philosophy of science, of morals, and of psychology, among others. It is possibly the single most influential work of philosophy by an Anglophone in the 19th century.

In the first excerpt, Mill agrees with the view of Whately that the term "species" means something different in natural history than in logic, and for the first time a clear label is applied to logical and natural species. However, he is rather vague and unclear on what makes differentiae within natural species not a marker of logical species, relying on some notion of them not being "primary", and due to extrinsic factors like climate or habits.

In the second he rejects Whewell's typical rather than total notion of natural kinds, setting off a debate over natural kinds that persists until the present day. Interestingly, though, he is not an essentialist; instead he states that the *kinds* are essentialist, but that the membership of things in that kind merely requires these things have a greater resemblance to items within it than those without it.

> According to [the schoolmen's] language, the proximate (or lowest) Kind to which any individual is referrible, is called its species. ... Man, therefore, we may call a species; Christian, or Mathematician, we cannot.
>
> Note here, that it is by no means intended to imply that there may not be different Kinds, or logical species, of Man. The various races and temperaments, the two sexes, and even the various ages, may be differences of kind, within our meaning of the term. ... classes are often mistaken for real Kinds, which are afterwards proved not to be so. But if it turned out that the differences were not capable of being thus accounted for [by a small number of primary differences], then Caucasian, Mongolian, Negro &c., would be really different Kinds of human beings, and entitled to be ranked as species by the logician, though not by the naturalist. For (as already noticed) the word species is used in a different signification in logic and in natural history. By the naturalist, organised beings are not usually said to be of different species, if it is supposed that they have descended from the same stock. That, however, is a sense artificially given to the word for the technical purposes of a particular science. To the logician, if a negro and a white man differ in the same manner (however less in degree) as a horse and a camel do, that is, if their differences are inexhaustible, and not referrible to any common cause, they are different species, whether they are descended from common ancestors or not. But if their differences can all be traced to climate and habits, or to some one or a few special differences in structure, they are not, in

the logician's view, specifically distinct. [*System of Logic*, 1843, Bk I, chap. 7, §4 (Mill 1974: 80f)

After the recognition and definition, then, of the *infimae species*, the next step is to arrange those *infimae species* into larger groups: making these groups correspond to Kinds wherever it is possible, but in most cases without any such guidance. And in doing this it is true that we are naturally and properly guided, in most cases at least, by resemblance to a type. We form our groups round certain selected Kinds, each of which serves as a sort of exemplar of its group. But though the groups are suggested by types, I cannot think that a group when formed is *determined* by the type; that in deciding whether a species belongs to the group, a reference is made to the type, and not to the characters; that the characters "cannot be expressed in words". This assertion is inconsistent with Dr. Whewell's own statement of the fundamental "principle" of classification, namely, that "general assertions shall be possible". If the class did not possess any characters in common, what general assertions would be possible respecting it? Except that they all resemble each other more than they resemble anything else, nothing whatever could be predicated of the class.

The truth is, on the contrary, that every genus or family is framed with distinct reference to certain characters, and is composed, first and principally, of species which agree in possessing all those characters. To these are added, as a sort of appendix, such other species, generally in small number, as possess *nearly* all the properties selected; wanting some of them one property, some another, and which, while they agree with the rest *almost* as much as these agree with one another, do not resemble in an equal degree any other group. Our conception of the class continues to be grounded on the characters; and the class might be defined, those things which *either* possess that set of characters, *or* resemble the things that do so, more than they resemble anything else.

And this resemblance itself is not, like resemblance between simple sensations, an ultimate fact, unsusceptible of analysis. Even the inferior degree of resemblance is created by the possession of common characters. Whatever resembles the genus Rose more than it resembles any other genus, does so because it possesses a greater number of the characters of that genus, than of the characters of any other genus. Nor can there be the smallest difficulty in representing, by an enumeration of characters, the nature and degree of resemblance which is strictly sufficient to include any object in the class. There are always some properties common to all things which are included. Others there often are, to which some things, which are nevertheless included, are exceptions. But the objects which are exceptions to one character are not exceptions to another: the resemblance which fails in some particulars must be made up for in others. The class, therefore, is constituted by the possession of *all* the characters which are universal, and *most* of those which admit of exceptions. [*System of Logic*, 1843, Bk IV, chap. 7, §4 (Mill 1974: 472)]

Jean Louis Rodolphe Agassiz (1807–1873)

Had there been no Darwin, Louis Agassiz might have been the colossus astride the mid century. But his views were remarkably conservative (Lurie 1960; Winsor 1991), and he was perhaps the last practising, respected, scientist to be a total fixist. In Agassiz' case it seems not merely to be piety, unlike Linnaeus and Ray, that makes him thus, but his attachment to Cuvier's work and ideas (he was a student of Cuvier) and in his case a literal Platonism. He thought that types and classes were thoughts in God's mind, and that actually the individual organisms did not properly instantiate these types. Not only was he a species realist (in the older Idealistic sense of "realist") he was a realist of all the Linnaean taxonomic ranks. He vehemently opposed a strawman version, of his own construction, of Darwin's ideas, but late in life became more open to them, though he never accepted they were correct.

> *… no so-termed character—that is, no observable mark—can be so striking as to indicate an absolute specific distinction; but at the same time, it should never be regarded as so trifling as to point to absolute identity; that characters do not mark off species, but that the combined relations to the external world in all circumstances of life do.* [italics original (Agassiz 1842)]

> There is a system in nature … to which the different systems of authors are successive approximations. … This growing co-incidence between our systems and that of nature shows … the identity of the operations of the human and the Divine intellect … [*Essay on Classification* (Agassiz 1859: 31; quoted in Lurie 1960)]

> Species then exist in nature in the same manner as any other groups, they are quite as ideal in the mode of existence as genera, families, etc., or quite as real. … Now as truly as individuals, while they exist, represent their species for the time being and do not constitute them, so truly do these same individuals represent at the same time their genus, their family, their order, their class, and their type, the characters of which they bear as indelibly as those of the species. [*Essay on Classification*, 1859 (quoted in Winsor 1979: 98)]

> … classification, rightly understood, means simply the creative plan of God as expressed in organic terms" [*Methods of Study in Natural History* (Agassiz 1863: 42)]

All the more comprehensive groups, equally with Species, are based upon a positive, permanent, specific principle, maintained generation after generation with all its essential characteristics. Individuals are the transient representatives of all these organic principles, which certainly have an independent, immaterial existence, since they outlive the individuals that embody them, and are no less real after the generation that has represented them for a time has passed away, than they were before" [*Op. cit.,* 136]

All the observations relative to domestic animals, among which there are so many and so numerous variations, again did not succeed in demonstrating a sufficiently large amplitude in these variations; never did they [the Darwinists] have as a result anything which manifests the indefinite tendency to a changeability without limit... [Chapter added to French edition of *Methods*, 1869 (Agassiz 1869; Morris 1997)]

[I]f species do not exist at all, as the supporters of the transmutation theory maintain, how can they vary? And if individuals alone exist, how can differences which may be observed among them prove the variability of species? [(in Lurie 1960: 297)]

James Dana (1813–1895)

Dana was an ally of Agassiz, until they had a falling out, and published a scientific journal for which he wrote himself. He was more Aristotelian in his metaphysics than Agassiz, and held that species had an immanent form, a potentiality, that causally made them what they were.

A species corresponds to a specific amount or condition of concentrated force, defined in the act or law of creation. [p306. Italics original (Dana 1857)]

The species is not the adult resultant of growth, nor the initial germ cell, nor its condition at any other point; it comprises the whole history of development. Each species has its own special mode of development as well as ultimate form or result, its serial unfolding, inworking and outflowing; so that the precise nature of the potentiality in each is expressed by the line that historical progress from the germ to the full expansion of its powers, and the realization of the end of its being. We comprehend the type-idea only when we understand the cycle of evolution [*sensu* development—JSW] through all its laws of progress, both as regards the living structure under development within, and its successive relations to the external world. [*Op. cit.* p308]

Richard Owen (1804–1892)

Owen is often unfairly treated by the historiography of the post-Darwinians as being a fixist, a conservative, and religious. All these are partly true, but it must be recalled that he also held that species could be transformed, and that he developed the notion of *Archetypes,* which Darwin employed as the basis for ancestral conditions, and of *homology,* on which all comparative biology now relies. In addition to this, he also named the first dinosaurs, and initially was fairly open to Darwin's views, although under the influence of Bishop Wilberforce and other religious authorities, and in reaction to the withering attacks of Huxley and others, he began to harden in his opposition (Camardi 2001).

> I apprehend that few naturalists now-a-days, in describing and proposing a name for what they call 'a new *species,*' use that term to signify what was meant by it twenty or thirty years ago, that is, an originally distinct creation, maintaining its primitive distinction by obstructive generative peculiarities. The proposer of the new species now intends to state no more than he actually knows; as for example, that the differences in which he founds the specific character are constant in individuals of both sexes, so far as observation has reached; and that they are not due to domestication or to artificially superinduced external circumstances, or to any outward influence within his cognizance; that the species is wild, or is such as it appears by nature. [*On the Osteology of the Chimpanzee and Orang Utan* (Owen 1835; quoted in Huxley 1906: 303)]
>
> No doubt the type-form of any species is that which is best adapted to the conditions under which such species at the time exists; and so long as those conditions remain unchanged, so long will the type remain; all varieties departing therefrom being in the same ratio less adapted to the evironing conditions of existence. But, if those conditions change, then the variety of the species at an antecedent date and state of things will become the type-form of the species at a later date, and in an altered state of things. [Presidental Address to the British Association for the Advancement of Science in 1858 (Basalla, Coleman, and Kargon 1970: 329)]

Philip Henry Gosse (1810–1888)

Gosse is best remembered for being the Father of the book *Father and Son* (Gosse 1970), depicted by his son as a harsh religious fanatic. But he was also a respected naturalist and is held to be the founder of the fashion for aquariums. He wrote several popular books before publishing his most infamous work,

Creation (Omphalos)[7] in which he argued that the universe and all life was indeed created as the book of Genesis described, but with the appearance of age, because all things were created in the middle of their lifecycle or other causal process, leading to the universe appearing old.

> Each order was distributed into subordinate groups, called Genera, and each genus into Species. As this last term is often somewhat vaguely used, it may not be useless to define its acceptation. It is used to signify those distinct forms which are believed to have proceeded direct from the creating hand of God, and on which was impressed a certain individuality, destined to pass down through all succeeding generations, without loss and without confusion. Thus the Horse and the Ass, the Tiger and the Leopard, the Goose and the Duck, though closely allied in form, are believed to have descended from no common parentage, however remote, but to have been primary forms of the original creation. It is often difficult in practice to determine the difference or identity of species; as we know of no fixed principle on which to found our decision, except the great law of nature, by which specific individuality is preserved—that the progeny of mixed species shall not be fertile *inter se*. [*An Introduction to Zoology* 1844 (p xv, cited in Simpson 1925: 175)]

> I demand also [as well as the creation of matter out of nothing—*JSW*], in opposition to the development hypothesis [pre-Darwinian evolutionism—*JSW*], the perpetuity of specific characters, from the moment when the respective creatures were called into being, till they cease to be. I assume that each organism which the Creator educed was stamped with an indelible specific character, which made it what it was, and distinguished it from everything else, however near or like. I assume that such character has been, and is, indelible and immutable; that the characters which distinguish species from species *now*, were as definite at the first instant of their creation as now, and are as distinct now as they were then. [*Creation (Omphalos)* 1857 (Gosse 1857: 111)][8]

Charles Lyell (1797–1875)

In his time, Lyell was not regarded as a significant figure in geology until late in his career, but he established a view that Whewell called "uniformitarianism", in which the causes of geological formations were held to be more or less constant

---

7   Also titled as *Omphalos*, the way it is called in *Father and Son*. My copy has the alternate title.

8   I am indebted, literally and metaphorically, to Dr Noelie Alito for purchasing an original copy of Gosse's *Creation (Omphalos)* on my behalf. A scanned copy is available on the internet at www.archive.org.

and the same today as in the past. His view influenced a young naturalist voyaging around the world in the *HMS Beagle.* Volume 2 of his *Principles of Geology* was received en route by Darwin, and it was a sustained treatment of the evolutionary views of Lamarck, including much discussion of the origin of species and what species were. These discussions no doubt triggered Darwin's thinking in many ways, although he seems to have annotated a later edition more than the one he received during his trip.

The name of species, observes Lamarck, has been usually applied to every collection of similar individuals, produced by other individuals like themselves. This definition, he admits, is correct, because every living individual bears a very close resemblance to those from which it springs. But this is not all which is usually implied by the term species, for the majority of naturalists agree with Linnaeus in supposing that all the individuals propagated from one stock have certain distinguishing characters in common which will never vary, and which have remained the same since the creation of each species. [*Principles of Geology* ((Lyell 1832, Vol. 2: 2)]

For the reasons, therefore, detailed in this and the two preceding chapters, we draw the following inferences, in regard to the reality of species in nature.

First, That there is a capacity in all species to accommodate themselves, to a certain extent, to a change of external circumstances, this extent varying greatly according to the species.

2dly. When the change of situation which they can endure is great, it is usually attended by some modifications of the form, colour, size, structure, or other particulars; but the mutations thus superinduced are governed by constant laws, and the capability of so varying forms part of the permanent specific character.

3dly. Some acquired peculiarities of form, structure, and instinct, are transmissible to the offspring; but these consist of such qualities and attributes only as are intimately related to the natural wants and propensities of the species.

4thly. The entire variation from the original type, which any given kind of change can produce, may usually be effected in a brief period of time, after which no farther deviation can be obtained by continuing to alter the circumstances, though ever so gradually,—indefinite divergence, either in the way of improvement or deterioration, being prevented, and the least possible excess beyond the defined limits being fatal to the existence of the individual.

5thly. The intermixture of distinct species is guarded against by the aversion of the individuals composing them to sexual union, or by the sterility of the mule offspring. It does not appear that true hybrid races have ever been perpetuated for several generations, even by the assistance of man; for the cases usually cited relate to the crossing of mules with individuals of pure species, and not to the intermixture of hybrid with hybrid.

6thly. From the above considerations, it appears that species have a real existence in nature, and that each was endowed, at the time of its creation, with the attributes and organization by which it is now distinguished. [*Principles of Geology* (Lyell 1832, Vol. 2: 64f)]

There are fixed limits beyond which the descendants from common parents can never deviate from a common type. ... It is idle ... to dispute about the abstract possibility of the conversion of one species into another, when there are known causes, so much more active in their nature, which must always intervene and prevent the actual accomplishment of such conversions. [From the 1835, 3rd, edition, vol II, p 162; the second sentence is not found in the first edition (quoted in Mayr 1982: 405)]

Augustin-Pyramus de Candolle (1778–1841)

De Candolle senior (the son, Alphonse, was a major figure in his own right in botany, who particularly emphasised variation in species) was a widely read thinker about the new botany. For example, he was read carefully by Darwin. Here, invariance of characters indicates species rank.

[A species is] ... the collection of all the individuals who resemble one another more than they resemble others; who are able, by reciprocal fecundation, to produce fertile individuals; and who reproduce by generation, such kind as one may by analogy suppose that all came down originally from one single individual. [*Théorie élementaire de la botanique* (Candolle 1819; quoted in Hunter Dupree 1968: 54)

By species we understand a number of plants, which agree with one another in invariable marks.

In this matter, everything depends on the idea of invariableness. When an organ, or a property of it, is changed neither by difference of soil, of climate, or of treatment, nor by continual breeding, this organ or property is said to be invariable. When, for instance, we have remarked during centuries, that the centrifolia has always unarmed leafstalks, we say correctly, that this property of centrifolia is invariable.

This idea proceeds on the supposition that the species which we know have existed as long as the earth has had its present form. No doubt there were, in the preceding state of our globe, other species of plants, which have now perished, and the remains of which we still find in impressions in shale, slate-clay, and other flœtz rocks. Whether the present species, which often resemble these, have arisen from them; whether the great revolutions on the surface of the earth, which we read in the Book of Nature, contributed to these transitions—we know not. What we know is that from as early a time as the human race has left memorials of its existence upon the earth the separate species of plants have maintained the same properties invariably.

To be sure, we frequently speak of the transitions and crossings of species; and it cannot be denied that something of this kind does occur, though without affecting the idea of species which we have proposed. We must, therefore, understand this difference.

We perceive the *Transitions of Species*, when it loses or changes the properties, which we had considered as invariable in the character. Thus, it would be a transition, if we had stated as an invariable character of winter wheat (*Triticum hybernum*), that it was biennial, and had an ear without awns; and if we should remark, that by frequent reproduction, and by very different treatment, it began to assume awns, and, when sown in spring, came to maturity during the same summer.

But this shows only that our idea of the difference between two kinds of grains had been incorrect; for it is the universal rule, that the character does not constitute the species, but the species the character. Species, then, only appear to undergo transitions, when we have considered an organ or property as invariable which is not so.

All properties of plants which are subject to change, form either a subspecies or a variety. By the former we understand such forms as continue indeed during some reproductions, but at last, by a greater difference of soil, of climate, and of treatment, are either lost or changed. When the different cabbage species receive the same treatment in the same climate, they continue to be frequently reproduced, without changing their appearance. But we can not on this account maintain, that cauliflower would retain the same favorite form in very different climates, and under a complete change of treatment. It at last changes so much, that it can scarcely be distinguished from the common cabbage. This, therefore, is a subspecies. Varieties again do not retain their forms during reproduction. The variable colors – the very variable taste, and other properties of the kitchen vegetables, the ornamental plants, and the fruit-trees, show what varieties are; and the scientific botanist must therefore be particularly attentive to distinguish permanent species from the variable subspecies, degenerate plants and varieties,

To this discrimination belongs, above all things, a careful, continued, and unprejudiced observation of the whole vegetation of the same plant during its different ages, and amidst the most different circumstances which have an influence on it. When, for instance, in the common *Lotus corniculatus*, on whatever soil it may grow, we uniformly observe that it has a solid stem, even and erect divisions of the calyx, and expanded filaments, we must of necessity distinguish, as a particular species from it, another form which grows in bogs and in watery meadows, which has a much higher, and always hollow stalk, the divisions of its calyx spread out into a star-shape and hairy, and which has uniformly thin filaments; and we must name this latter species either *Lotus uliginosus* with Schkuhr, or *Lotus major* with Scopoli and Smith. As, on the other hand, the *Pimpinella Saxifraga* grows sometimes quite smooth, and sometimes in woods and shady meadows, considerably hairy; as it displays sometimes simple and small stem-leaves, sometimes half and even doubly pinnated leaves; and as these forms vary according to the situation of the plant and during reproduction, we can not regard these forms by any means as distinct species, but we must view them as corruptions.

We see, that, in order to decide respecting the idea of a species, an observation of many years, and of much accuracy, is often required; and that the cultivation of plants, from the most different climates, in botanical gardens, is in the highest degree necessary for their discrimination. [*Elements of the Philosophy of Plants* 1813 Fr. edn (Candolle and Sprengel 1821; quoted in Britton 1908: 228ff)]

John Lindley (1799–1865)

Lindley was a follower of Jussieu's approach to classification, rather than Linnaeus'. A botanist, his *Introduction* was widely read by the new genetlemen botanical enthusiasts. He here proposes both something like the reproductive isolation test of species, and the criticism that this is impractical, which was later raised against Mayr's definition. His distinction between "natural species" which are reproductively interfertile, and "botanical species" which are those based on fructification, indicates that he didn't think Linnaean species were much more than conveniences.

A *species* is a union of individuals agreeing with each other in all essential characters of vegetation and fructification, capable of reproduction by seed without change, breeding freely together, and producing perfect seed from which a fertile progeny can be reared. Such are the true limits of a species; and if it were possible to try all plants by such a test, there would be no difficulty in fixing them, and determining what is species and what is variety. But, unfortunately, such is not the case. The manner in which individuals agree in their external characters is the only guide which can be followed in time greater part of plants.

We do not often possess the means of ascertaining what the effect of sowing their seed or mixing the pollen of individuals would be; and, consequently, this test, which is the only sure one, is, in practice, seldom capable of being applied. The determination of what is a species, and what a variety, becomes therefore wholly dependent upon external characters, the power of duly appreciating which, as indicative of specific difference, is only to be obtained by experience, and is, in all cases, to a certain degree arbitrary. It is probable that, in the beginning, species only were formed; and that they have, since the creation, sported into varieties, by which time limits of the species themselves have now become greatly confounded. For example, it may he supposed that a rose, or a few species of rose, were originally created. In the course of time these have produced endless varieties, some of which, depending for a long series of ages upon permanent peculiarities of soil or climate, have been in a manner fixed, acquiring a constitution and physiognomy of their own. Such supposed varieties have again intermixed with each other, producing other forms, and so the operation has proceeded. But as it is impossible, at the present day, to determine which was the original or originals, from which all the roses of our own time have proceeded, or even whether they were produced in the manner I have assumed; and as the forms into which they divide are so peculiar as to render a classification of them indispensable to accuracy of language; it has become necessary to give names to certain of those forms, which are called species. Thus it seems that there are two sorts of species: the one, called natural species, determined by the definition given above; and the other, called botanical species, depending only upon the external character of the plant. The former have been ascertained to a very limited extent; of the latter nearly the whole of systematic botany consists. In this sense a species may be defined to be "all assemblage of individuals agreeing in all the essential characters of vegetation and fructification." Here time whole question lies with the word essential. What is an essential character of a species? This will generally depend upon a. proneness to vary, or to be constant in particular characters, so that one class of characters may be essential in one genus, another class in another genus; and these points can be only determined by experience. Thus, in the genus Dahlia, the form of the leaves is found to be subject to great variation; the same species producing from seed, individuals, the forum of whose leaves vary in a very striking manner; the form of the leaves is, therefore, in Dahlia, not a specific character. In like manner, in Rosa, the number of prickles, the surface of the fruit, or the surface of their leaves, and their serratures, are found to be generally fluctuating characters, and can not often be taken as essential to species. The determination of species is, therefore, in all respects, arbitrary, and must depend upon the discretion or experience of the botanist. [*An Introduction to Botany* (Lindley 1832; quoted in Britton 1908: 231f)]

633. It seems to be generally admitted by those who have turned their attention to the consideration of the manner in which organized beings are related to each other, that each species is allied to many others in different degrees, and that such relationship is best expressed by rays (the affinities) proceeding from a common centre (the species). In like manner, in studying the mutual relationship of the several parts of the vegetable kingdom, the same form of distribution constantly forces itself upon the mind; genera and orders being found to be apparently the centre of spheres, whose surface is only defined by the points where the last traces of affinity disappear.

...

638. What we call the characters of plants are merely the signs by which we judge of affinity, and all the groups into which plants are thrown are in one sense artificial, inasmuch as nature recognises no such groups.

639. Nevertheless, consisting in all cases of species very closely allied in nature, they are in another sense natural.

640. But as the classes, subclasses, groups, alliances, natural orders, and genera of botanists have no real existence in nature, it follows that they have no fixed limits, and consequently that it is impossible to define them.

641. They are to be considered as nothing more than the expression of particular *tendencies* (nixus), on the part of the plants they comprehend, to assume a particular mode of developement.

642. Their characters are therefore nothing more than a declaration of their prevailing tendencies, and are liable to numerous exceptions. ...

644. The following propositions seem incontrovertible:

1. Nothing that is constant can be regarded as unimportant.

2. Every thing constant must be dependent upon or connected with some essential function. Therefore all constant characters, of whatever nature, require to be taken into account in classifying plants according to their natural affinities. [*Elements of Botany, Structural, Physiological, Systematical, and Medical* (1841: 85–88)]

Robert Chambers (1802–1871)

Chambers published *Vestiges* anonymously, and it caused a storm over its transformism. Darwin was in part discouraged by the reaction of the scientific community to it, and held off publishing until famously he was forced to by the arrival of Wallace's letter. Oddly, Chambers does not hold any unusual notion of what species are.

> … species, the subdivision where intermarriage or breeding is usually considered as natural to animals, and where a resemblance of offspring to parents is generally persevered in. [*Vestiges of the History of Creation* (Chambers 1844: 263)]

Prince Charles Lucien Bonaparte (1803–1857)

Bonaparte is the nephew of Napoleon, but he was an ornithologist rather than a political figure, despite being detained by the British until the Napoleonic Wars were at an end. His view of species was transformist and he held that the environment was the cause of their stability.

> We will state with unanimous conviction that the antediluvian crocodiles, elephants and rhinoceroses were the ancestors of those living in our day, and these animals would not have been able to continue to exist without the manifold mutations that their systems produced to adapt themselves to the environment, and that became second nature to their descendants. … If the environment remains the same, so do the species. The stabilizing influence is then by itself all-powerful. The mutating influence can succeed in opposing it only when the whole world surrounding it changes. … But races, however different in characteristics they may be, vanish entirely or at least do not long survive as soon as the environment that produced them ceases to be the same … The transitions between the different races and their type are the best evidence that we can supply to set aside putative species, which are to be relegated to races, with which the painstaking zoologist must nevertheless occupy himself just as earnestly. [(Quoted in Stresemann 1975: 166)]

Hermann Schaffenhausen (1816–1893)

> The immutability of species which most scientists regard as natural law is not proved, for there are no definite and unchangeable characteristics of the species, and the borderline between species and subspecies [*Art* and *Abart*] is wavering and uncertain. ["On the Constancy and Transformation of Species" 1853 (quoted in Temkin 1959: 342)]

Joseph Dalton Hooker (1817–1911)

Hooker as a botanist was one of Darwin's earliest and closest confidants, and became the Director of the Royal Botanic Gardens at Kew.

Bother variation, developement [sic] & all such subjects,! It is reasoning in a circle after all. As a Botanist I must be content to take species as *they appear to be* not as *they are*, **& still less** as **they were** or ought to be. [To Darwin, July 1845 (Stevens 1997: 349)]

It is very much to be wished that the local botanist should commence his studies upon a diametrically opposite principle to that upon which he now proceeds, and that he should endeavour, by selecting good suites of specimens, to determine *how few*, not *how many* species are comprised in the flora of his district. The permanent differences will, he may depend upon it, soon force themselves upon his attention, whilst those which are non-essential will consecutively be eliminated. There is no better way of proving the validity of characters than by attempting to invalidate them. [*Botany of the Antarctic Voyage* volume 2 (Stevens 1997: 364n62)]

Isidore Geoffroy Saint Hilaire (1805–1861)

Son of Etienne Geoffroy Saint Hilaire, Isidore was a widely respected zoologist specialising in sports and deviants, and one of Darwin's favoured naturalist writers. In his *General Natural History*, he has one of the first broad discussions of definitions of species.

The species is the fundamental group given by nature. Everything comes from it or returns to it; like the variety which is an accidental derivation, and race which is a derivation become permanent; and like the family or "company", the company, the aggregate and the community, which are natural subdivisions [of species]; like the genus which is the collection of the species which most closely resemble each other; and like the higher groups themselves which are collections of genera, and consequently, mediately, of species.1 [Chapitre XXXVI "Définitions diverses de l'espèce organique et résumé des vues émises sur les rapports des êtres actuels avec ceux des temps antérieurs" (Geoffroy Saint Hilaire 1859: 365f). Thanks to Laurent Penet for help with the translation. Note that the rank of "family" was not at this time considered a standard rank in the Linnaean taxonomy, and terms like "company" were often used instead.]

---

1 L' espèce est le groupe fondamental donné par la nature. Tout en part ou y aboutit; comme la variété qui en est une dérivation accidentelle, et la race une dérivation devenue permanente; comme la famille ou compagnie, la société, l' agrégat et la communauté, qui en sont des subdivisions naturelles; comme le genre qui est la collection des espèces qui se ressemblent le plus; comme les groupes supérieurs eux-mêmes qui sont des collections de genres, par conséquent, médiatement, d' espèces.

George Henry Lewes (1817–1878)

Lewes was a bit of a renaissance man, publishing literature, literary criticism, philosophy, particularly of mind, and, as here, speculations in science. He was part of Darwin's social circle, and a friend of Erasmus Darwin, as well as of Mill and Carlyle, and he published the *Westminster Review*, from which this article is taken. It was published anonymously, but was excerpted in his *The Physiology of Common Life* in 1860 (Lewes 1860). In this article, Lewes argues that inheritance comes via an epigenetic mechanism provided by each parent, but which is sometimes masked by one or the other parent's contribution. He also discussed "atavisms" which were inherited from more distal ancestors such as grandparents.

Lewes proposes that species are Lockean conventions, fictions, not things. He says that organisms reproduce their own type, not the species', and that fixity of species *and* transmutation of species are both wrong, since there are none. Elsewhere in the article, he argues that attempts to regulate marriage of those who inherit tendencies to insanity and other diseases are misplaced due to our ignorance.

The article is interesting for the views immediately before Darwin's *Origin*, and for his attempt to be critical of the anecdotal evidence about heredity (something which Darwin himself was prone to accept too easily), although he still accepts that previous pregnancies can affect later ones.

Heritage (*l'hérédité*), or the transmission of physical and mental qualities from parents to offspring, is one of those general facts of Nature which lie patent to universal observation. Children resemble their parents. Were this law not constant, there could be no constancy of Species; the horse might engender an elephant, the squirrel might be the progeny of a lioness, the tadpole of a tapir. The law, however, is constant. During thousands of years the offspring has continued to exhibit the structure, the instincts, and all the characteristics of the parents. Every day some one exclaims—as if the fact surprised him—"That boy is the very image of his father!" yet no one exclaims, "How like that pug dog is to its parent!" Boys or pug dogs, all children resemble their parents. We do not allude to the fact out of any abstract predilection for truisms, but simply to marshal into due prominence an important truth, on which the whole discussion of heritage must rest. The truth is this: Constancy in the transmission of structure and character from parent to offspring, is a law of Nature.

That this truth is not a truism we shall show by at once contradicting, or at least qualifying it. The very same experience [137] which guarantees the constancy, also teaches, and with almost equal emphasis, that this constancy is not absolute. Variations occur. Children sometimes do *not* resemble their parents; which accounts for the exclamation of surprise when they do resemble them. Nay, the children are sometimes not only unlike their parents, they are, in important characteristics, unlike their Species. We then call them Deformities or Monsters, because, while their Species is distinguished by having four legs, they themselves have six or none; while their Species possesses a complex brain, they are brainless, or have imperfect brains; while their Species is known by its cloven hoof, they have solid hoofs, and so on.* Dissemblances as great are observable in moral characteristics. We see animals of ordinary aptitudes engender offspring sometimes remarkable for their fine qualities, and sometimes for their imbecility. The savage wolf brings forth occasionally a docile, amiable cub; the man of genius owns a blockhead for his son. In the same family we observe striking differences in stature, aspect, and disposition. Brothers brought up together in the same nursery, and under the same tutor, will differ as much from each other as they differ from the first person they meet. From Cain and Abel down to the brothers Bonaparte, the striking opposition of characters in families has been a theme for rhetoric. Nor is this all. In cases where the consanguinity may be said to be so much nearer than that of ordinary brotherhood, namely, in twins, we see the same diversity; and this diversity is exhibited in those rare cases where the twins have *only one body between them.* The celebrated twins Rita and Christina† were so *fused* together, that they had only two legs between them: two legs and four arms and two heads; yet they were quite different in disposition. The same difference was manifested in the celebrated Presburg twins, and in the African twins recently exhibited in London.

* "Flachsland rapporte quo deux époux bien constitutés mirent an mothe trois en- fans sans avantbras ni jambes; d'autres dont parle Schmucker n'eurent que des enfans munis de douze orteils at douze doights."—Burdach, *Traité de Physiologie* ii. 264.

† See Geoffroy St. Hilaire, "Philosophie Anatomique," vol. ii.; and Serres, "Recherches d'Anatomie Transcendante."

It is clear, then, that offspring do not always closely resemble parents; and it is further clear from the diversities in families, that they do not resemble them in equal degrees. Two brothers may be very unlike each other, and yet both like their parents; but the resemblance to the parents must, in this case, be variable. So that when we lay down the rule of *constancy in transmission*, we must put a rider on it, to the effect that this Constancy is not absolute, but is accompanied by a law of Variation. It is the [138] intervention of this law which makes hereditary influence a problem; without it, heritage would be as absolute as the union of acids with bases.

Some philosophers have tried to explain the law of constancy in transmission, and its independence of the law of variations, by maintaining that it is the

Species only, not the Individual, which is reproduced. Thus a sheep is always and everywhere a sheep, a man a man, reproducing the *specific type* but not necessarily reproducing any individual peculiarities. All sheep resemble each other, and all men resemble each other, because they all belong to specific types. What does the reader say to this hypothesis? Burdach, who adopts it,* adduces his facts: for example, a dog from whom the spleen was extirpated reproduced dogs with perfect spleens; an otter, deprived of its fore paws, produced six young with their legs quite perfect; in a word, "l'idée de l'espèce se reproduit dans le fruit et lui donna des organes qui manquaient au père ou I la mère." The hypothesis has seemed convincing to the majority of thinkers, but it labours under one fatal objection—namely, Species cannot reproduce itself; for Species does not exist. It is an entity, an abstract idea, not a concrete fact. It is a fiction of the understanding, not an object existing in Nature. The *thing* Species no more exists than the thing Goodness or the thing Whiteness. Nature only knows individuals. A collection of individuals so closely resembling each other as all sheep resemble each other, are conveniently classed under one general term, named species; but this general term has no objective existence; the abstract or typical sheep, apart from all concrete individuals, has no existence out of our systems. Whenever an individual sheep is born, it is the offspring of two individual sheep, whose structures and dispositions it reproduces; it is not the offspring of an abstract idea; it does not come into being at the bidding of a Type, which as a Species sits apart, regulating ovine phenomena. The facts of dissemblance between offspring and parents we shall explain by-and-by; they do not plead in favour of Species, because Species is a figment of philosophy, not a fact. The sooner we disengage our Zoology from all such lingering remains of old Metaphysics the better. Nothing but dreary confusion and word-splitting can come of our admitting them. Think of the hot and unwise controversies respecting "transmutation of species," which would have been spared if a clear conception of the meaning of Species had been steadily held before the disputants, or if the laws which regulate heritage had been duly considered. In one sense, transmutation of [139] Species is a contradiction in terms. To ask if one species can produce another; *i. e.*, a cat produce a monkey is to ask if the offspring do not inherit the organization of their parents. We know they do. We cannot conceive it otherwise. But the laws of heritage place the dispute in something of a clearer light, for they teach us that "Species" is constant, because individuals reproduce individuals closely resembling them, which is the meaning of "Species;" and they also teach us, that individuals reproduce individuals *varying* in structure from themselves, which Varieties, becoming transmitted as part and parcel of the parental influence, will, in time, become so great as to constitute a difference in Species. It is in vain that the upholders of a "fixity of Species" assert that all the varieties observed are differences of *degree* only. Differences of degree become differences of kind, when the gap is widened: ice and steam are only differences of degree, but they are equivalent to differences of kind. If, therefore, "transmutation of Species" is absurd, "fixity of Species" is not a whit less so. That which does not exist, can neither be transmuted nor maintained in fixity. Only individuals exist; they resemble their parents, and they differ from their parents. Out of these resemblances we create Species, out of these differences we create Varieties; we do so as conveniences of classification, and then believe in the reality of our own figments.

 * "Physiologie," ii. 245.

"Les espèces," said Buffon, boldly, "sont les seuls êtres de la nature,"[2] and thousands have firmly believed this absurdity. The very latest work published on this subject,* reproduces the dictum, and elaborately endeavours to demonstrate it. "Les espèces sont les formes primitives de la nature. Les individus n'en sont que des reprèsentations, des copies."[3] This was very well for Plato; but for a biologist of the nineteenth century to hold such language shows a want of philosophic culture. A cursory survey of the facts should have shown the error of the conception, if nothing else would. Facts plainly tell us that the individual and the individual's peculiarities, not those of the abstract Type, are transmitted.

 * "Cours de Physiologie Comparèe," par M. Flourens. 1856. A feeble and inaccurate book.

[(Lewes 1856: 136–139)]

---

2 "Species are the only being in nature."

3 "Species are the primitive forms in nature. Individuals are only representations, copies."

🎝 Section 3. Modern Conceptions

Charles Darwin is important not so much for the novelties on the nature of the species concept that he provided—there are only really two of these, failure to breed in nature, and selection as the motive force of specific characters. Rather it is because his book *On the Origin of Species* changed *every* scientist's way of looking at species thereafter. He has been more closely scrutinized than anybody else, and there is a wealth of material available. One thing that we should put out of our minds from the beginning, though: it is *not* true that Darwin did not address the origin of species in *On the Origin of Species* (cf. the epigrams). The book is "one long argument" (Mayr 1991) on that very point. Over and again, he discusses why species evolve to be distinct from parental forms, and how they have done so.

Charles Robert Darwin (1809–1882)

Few scientists have been so intensively scrutinised as Charles Darwin. In part this is because, as is obvious, he represents the commencement of a major revolution in thinking about humanity and the living world, but his novelty in matters of detail ought not to be oversold. He was not the first in any particular claim, but in the way he employed all the different views as a coherent system. On species, it is interesting to see him move from a reproductive conception of species to an adaptive one, but at all times, Darwin seems to have held species to be real objects in nature, despite the apparent nominalism of one comment in the *Origin* that has been taken as his final view, wrongly.

It is occasionally stated that Darwin denied the reality of species, or held that "species" was an arbitrary concept. Mayr notes "one might get the impression [from the *Origin of Species*] that he considered species as something purely arbitrary and invented merely for the convenience of taxonomists" (Mayr 1982: 268). Mayr goes on to note that he nevertheless treated species in a perfectly orthodox taxonomic manner and that he treated the concept purely typologically. Beatty (1985) and Ereshefsky (1999) raise similar doubts on Darwin's view of species. On the contrary, Darwin was a species realist, and although he developed his views over time he never ceased being a species realist, just as his mentor Lyell was.

Darwin has often been thought to present a "population thinking" approach to biology, and he does indeed do this, but he was not the first by far to think that species are variable, nor that individuals in a species each have their own unique typology.

It is daily happening, that naturalists describe animals as species … There is only two ways [*sic*] of proving to them it is not; one where they can [be] proved descendant [Kottler interpolates: descent from common parents], which of course most rare, or when placed together they will breed. [Notebook B122 (quoted in Kottler 1978)]

As species is real thing with respect to contemporaries—fertility must settle it [C152]

If they [systematists—*JSW*] give up infertility in largest sense as test of species—they must deny species which is absurd. [E24].

Definition of species: one that remains at large with constant characters, together with other beings of very near structure. [B213 (de Beer 1960: 66)]

It has long appeared to me, that the root of the difficulty in settling such questions as yours,—whether the number of species &c &c should enter as an element in settling the value of existence of a group—lies in our ignorance of what we are searching after in our natural classifications.—Linnaeus confesses profound ignorance.—Most authors say it is an endeavour to discover the laws according to which the Creator has willed to produce organized beings—But what empty high-sounding sentences these are—it does not mean order in time of creation, nor propinquity to any one type, as man.—in fact it means just nothing—According to my opinion, (which I give everyone leave to hoot at, like I should have, six years since, hooted at them, for holding like views) classification consists in grouping beings according to their actual *relationship*, ie, their consanguinity, or descent from common stocks … [Letter to G. R. Woodhouse, 26 July 1843, (Burkhardt 1996: 76)]

I was so struck with the distributions of Galapagos organisms &c &c & with the character of the American fossil mammifers, &c &c that I determined to collect blindly every sort of fact, which c^d bear any way on what are species … At last gleams of light have come, & I am almost convinced (quite contrary to the opinion I started with) that species are not (it is like confessing a murder) immutable. [Letter to Hooker *op. cit.* p80, 11 January 1844]

I thank you much for attempting to mark the list of dubious species: I was afraid it was a very difficult task, from, as you say, the want of a definition of what a species is.—I think however you were marking exactly what I wanted to know. My wish was derived as follows: I have ascertained, that APPARENTLY (I will not take up time by showing how) there is more variation, a wider geographical range, & probably more individuals, in the species of *large* genera than in the species of *small* genera. These general facts seem to me very curious, & I wanted to ascertain one point more; viz whether the closely allied and dubious forms which are generally considered as species, also belonged on average to large genera. [Letter to Henslow, 21 July 1855 (Barlow 1967: 182)]

You speak of species not having any material base to rest on; but is this any greater hardship than deciding what deserves to be called a variety & be designated by a greek letter. When I was at systematic work, I know I longed to have no other difficulty (great enough) than deciding whether the form was distinct enough to deserve a name; & not to be haunted with undefined & unanswerable question whether it was a true species. What a jump it is from a well marked variety, produced by natural cause, to a species produced by the separate act of the Hand of God. But I am running on foolishly.—By the way I met the other day Phillips, the Palaeontologist, & he asked me "how do you define a species?"—I answered "I cannot" Whereupon he said "at last I have found out the only true definition,—'any form which has ever had a specific name'! ... [letter to Asa Gray, 29 November 1857 (Burkhardt 1996: 183)]

I knew, of course, of the Cuvierian view of Classification, but I think that most naturalists look for something further, & search for 'the natural system',—'for the plan on which the Creator has worked' &c &c.—It is this further element which I believe to be simply genealogical. [Darwin to Huxley 3 October 1853 (Padian 1999: 355)]

Of the latter [rabbit, a piebald hybrid of black and gray breeds] I now possess a specimen, and it is marked about the head differently from the French specific description. This circumstance shows how cautious naturalists should be in making species; for even Cuvier, on looking at the skull of one of these rabbits, thought it was probably distinct! [*Journal of Researches* (Darwin 1839, Chapter IX: 184)]

The distinction of the rabbit as a species, is taken from peculiarities in the fur, from the shape of the head, and from the shortness of the ears. I may here observe that the difference between the Irish and English hare rests upon nearly similar characters, only more strongly marked. [*loc. cit.* p184n]

I was much struck with the marked difference between the vegetation of these eastern valleys and those on the Chilian side: yet the climate, as well as the kind of soil, is nearly the same, and the difference of longitude very trifling. The same remark holds good with the quadrupeds, and in a lesser degree with the birds and insects. I may instance the mice, of which I obtained thirteen species on the shores of the Atlantic, and five on the Pacific, and not one of them is identical. We must except all those species, which habitually or occasionally frequent elevated mountains; and certain birds, which range as far south as the Strait of Magellan. This fact is in perfect accordance with the geological history of the Andes; for these mountains have existed as a great barrier since the present races of animals have appeared; and therefore, unless we suppose the same species to have been created in two different places, we ought not to expect any closer similarity between the organic beings on the opposite sides of the Andes than on the opposite shores of the ocean. In both cases, we must leave out of the question those kinds which have been able to cross the barrier, whether of solid rock or salt-water.[5]

[5] This is merely an illustration of the admirable laws, first laid down by Mr. Lyell, on the geographical distribution of animals, as influenced by geological changes. The whole reasoning, of course, is founded on the assumption of the immutability of species; otherwise the difference in the species in the two regions might be considered as superinduced during a length of time.

[*Loc. cit.* Chapter XV, p313]

In those cases in which a genus includes only a single species, I have followed the practice of some botanists, and given only the generic character, believing it to be impossible, before a second species is discovered, to know which characters will prove of specific, in contradistinction to generic, value.

In accordance with the Rules of the British Association, I have faithfully endeavoured to give to each species the first name attached to it, subsequently to the introduction of the binomial system, in 1758, in the tenth edition.[1] In accordance with the Rules, I have rejected all names before this date, and all MS. names. In one single instance, for reasons fully assigned in the proper place, I have broken through the great law of priority. I have given much fewer synonyms than is usual in conchological works; this partly arises from my conviction that giving references to works, in which there is not any original matter, or in which the Plates are not of a high order of excellence, is absolutely injurious to the progress of natural history, and partly, from the impossibility of feeling certain to which species the short descriptions given in most works are applicable;—thus, to take the commonest species, the *Lepas anatifera*, I have not found a single description (with the exception of the anatomical description by M. Martin St. Ange) by which this species can be certainly discriminated from the almost equally common *Lepas Hillii*. I have, however, been fortunate in having been permitted to examine a considerable number of authentically named specimens, (to which I have attached the sign (!) used by botanists,) so that several of my synonyms are certainly correct.

> [1] In the Rules published by the British Association, the 12th edition, (1766,) is specified, but I am informed by Mr. Strickland that this is an error, and that the binomial method was followed in the 10th edition. of the 'Systema Naturæ.'

[Preface to his *A monograph on the sub-class Cirripedia* (Darwin 1851)]

… in my own work, I have not felt conscious that disbelieving in the *permanence* of species has made much difference one way or the other; in some few cases (if publishing avowedly on doctrine on non-permanence) I sh^d. *not* have affixed names, & in some few cases sh^d. have affixed names to remarkable varieties. Certainly I have felt it humiliating, discussing & doubting & examining over & over again, when in my own mind, the only doubt has been, whether the forms varied *today or yesterday* (to put a fine point on it, as Snagsby would say). After describing a set of forms, as distinct species, tearing up my M.S., & then making them one again (which has happened to me) I have gnashed my teeth, cursed species, & asked what sin I had committed to be so punished: But I must confess, that perhaps the same thing w^d. have happened to me on any scheme of work—… [Darwin to Hooker, 25 September 1853 (Burkhardt 1996: 128–129)]

The Origin of Species (1869, 5th edition used)

Indefinite variability is a much more common result of changed conditions than definite variability, and has probably played a more important part in the formation of our domestic races. We see indefinite variability in the endless slight peculiarities which distinguish the individuals of the same species, and which cannot be accounted for by inheritance from either parent or from some more remote ancestor. [p16]

Altogether at least a score of pigeons might be chosen, which, if shown to an ornithologist, and he were told that they were wild birds, would certainly be ranked by him as well-defined species. [p25]

May not those naturalists who, knowing far less of the laws of inheritance than does the breeder, and knowing no more than he does of the intermediate links in the long lines of descent, yet admit that many of our domestic races are descended from the same parents—may they not learn a lesson of caution, when they deride the idea of species in a state of nature being lineal descendants of other species? [p29]

But what concerns us is that the domestic varieties of the same species differ from each other in almost every character, which man has attended to and selected, more than do the distinct species of the same genera. [p37]

Before applying the principles arrived at in the last chapter to organic beings in a state of nature, we must briefly discuss whether these latter are subject to any variation. To treat this subject properly, a long catalogue of dry facts ought to be given; but these I shall reserve for a future work. Nor shall I here discuss the various definitions which have been given of the term species. No one definition has satisfied all naturalists; yet every naturalist knows vaguely what he means when he speaks of a species. Generally the term includes the unknown element of a distinct act of creation. The term "variety" is almost equally difficult to define; but here community of descent is almost universally implied, though it can rarely be proved. We have also what are called monstrosities; but they graduate into varieties. By a monstrosity I presume is meant some considerable deviation of structure, generally injurious, or not useful to the species. Some authors use the term "variation" in a technical sense, as implying a modification directly due to the physical conditions of life; and "variations" in this sense are supposed not to be inherited; but who can say that the dwarfed condition of shells in the brackish waters of the Baltic, or dwarfed plants on Alpine summits, or the thicker fur of an animal from far northwards, would not in some cases be inherited for at least a few generations? And in this case I presume that the form would be called a variety. [p38]

There is one point connected with individual differences, which is extremely perplexing: I refer to those genera which have been called "protean" or "polymorphic," in which the species present an inordinate amount of variation. With respect to many of these forms, hardly two naturalists agree whether to rank them as species or as varieties. We may instance Rubus, Rosa, and Hieracium amongst plants, several genera of insects and of Brachiopod shells. In most polymorphic genera some of the species have fixed and definite characters. Genera which are polymorphic in one country, seem to be, with a few exceptions, polymorphic in other countries, and likewise, judging from Brachiopod shells, at former periods of time. These facts are very perplexing, for they seem to show that this kind of variability is independent of the conditions of life. I am inclined to suspect that we see, at least in some of these polymorphic genera, variations which are of no service or disservice to the species, and which consequently have not been seized on and rendered definite by natural selection, as hereafter to be explained. [p40]

The forms which possess in some considerable degree the character of species, but which are so closely similar to other forms, or are so closely linked to them by intermediate gradations, that naturalists do not like to rank them as distinct species, are in several respects the most important for us. We have every reason to believe that many of these doubtful and closely allied forms have permanently retained their characters for a long time; for as long, as far as we know, as have good and true species. Practically, when a naturalist can unite by means of intermediate links any two forms, he treats the one as a variety of the other; ranking the most common, but sometimes the one first described, as the species, and the other as the variety. But cases of great difficulty, which I will not here enumerate, sometimes arise in deciding whether or not to rank one form as a variety of another, even when they are closely connected by intermediate links; nor will the commonly assumed hybrid nature of the intermediate forms always remove the difficulty. In very many cases, however, one form is ranked as a variety of another, not because the intermediate links have actually been found, but because analogy leads the observer to suppose either that they do now somewhere exist, or may formerly have existed; and here a wide door for the entry of doubt and conjecture is opened.

Hence, in determining whether a form should be ranked as a species or a variety, the opinion of naturalists having sound judgment and wide experience seems the only guide to follow. We must, however, in many cases, decide by a majority of naturalists, for few well-marked and well-known varieties can be named which have not been ranked as species by at least some competent judges. [p41]

The geographical races or sub-species are local forms completely fixed and isolated; but as they do not differ from each other by strongly marked and important characters, "There is no possible test but individual opinion to determine which of them shall be considered as species and which as varieties." [Quoting Wallace] Lastly, representative species fill the same place in the natural economy of each island as do the local forms and sub-species; but as they are distinguished from each other by a greater amount of difference than that between the local forms and sub-species, they are almost universally ranked by naturalists as true species. Nevertheless, no certain criterion can possibly be given by which variable forms, local forms, sub-species, and representative species can be recognised. [p42]

Some few naturalists maintain that animals never present varieties; but then these same naturalists rank the slightest difference as of specific value; and when the same identical form is met with in two distinct countries, or in two geological formations, they believe that two distinct species are hidden under the same dress. The term species thus comes to be a mere useless abstraction, implying and assuming a separate act of creation. It is certain that many forms, considered by highly competent judges to be varieties, resemble species so completely in character, that they have been thus ranked by other highly competent judges. But to discuss whether they ought to be called species or varieties, before any definition of these terms has been generally accepted, is vainly to beat the air. [p43]

I may here allude to a remarkable memoir lately published by A. de Candolle, on the oaks of the whole world. No one ever had more ample materials for the discrimination of the species, or could have worked on them with more zeal and sagacity. ... De Candolle then goes on to say that he gives the rank of species to the forms that differ by characters never varying on the same tree, and never found connected by intermediate states. After this discussion, the result of so much labour, he emphatically remarks: "They are mistaken, who repeat that the greater part of our species are clearly limited, and that the doubtful species are in a feeble minority. This seemed to be true, so long as a genus was imperfectly known, and its species were founded upon a few specimens, that is to say, were provisional. Just as we come to know them better, intermediate forms flow in, and doubts as to specific limits augment." [p43f]

When a young naturalist commences the study of a group of organisms quite unknown to him, he is at first much perplexed in determining what differences to consider as specific, and what as varietal; for he knows nothing of the amount and kind of variation to which the group is subject; and this shows, at least, how very generally there is some variation. But if he confine his attention to one class within one country, he will soon make up his mind how to rank most of the doubtful forms. His general tendency will be to make many species, for he will become impressed, just like the pigeon or poultry fancier before alluded to, with the amount of difference in the forms which he is continually studying; and he has little general knowledge of analogical variation in other groups and in other countries, by which to correct his first impressions. As he extends the range of his observations, he will meet with more cases of difficulty; for he will encounter a greater number of closely allied forms. But if his observations be widely extended, he will in the end generally be able to make up his own mind; but he will succeed in this at the expense of admitting much variation, and the truth of this admission will often be disputed by other naturalists. When he comes to study allied forms brought from countries not now continuous, in which case he cannot hope to find intermediate links, he will be compelled to trust almost entirely to analogy, and his difficulties will rise to a climax.

Certainly no clear line of demarcation has as yet been drawn between species and sub-species—that is, the forms which in the opinion of some naturalists come very near to, but do not quite arrive at, the rank of species: or, again, between sub-species and well-marked varieties, or between lesser varieties and individual differences. These differences blend into each other by an insensible series; and a series impresses the mind with the idea of an actual passage. [p44f]

From these remarks it will be seen that I look at the term species as one arbitrarily given, for the sake of convenience, to a set of individuals closely resembling each other, and that it does not essentially differ from the term variety, which is given to less distinct and more fluctuating forms. The term variety, again, in comparison with mere individual differences, is also applied arbitrarily, for convenience' sake. [p46]

From looking at species as only strongly marked and well-defined varieties, I was led to anticipate that the species of the larger genera in each country would oftener present varieties, than the species of the smaller genera; for wherever many closely related species (i.e., species of the same genus) have been formed, many varieties or incipient species ought, as a general rule, to be now forming. Where many large trees grow, we expect to find saplings. Where many species of a genus have been formed through variation, circumstances have been favourable for variation; and hence we might expect that the circumstances would generally be still favourable to variation. On the other hand, if we look at each species as a special act of creation, there is no apparent reason why more varieties should occur in a group having many species, than in one having few. [p47]

Moreover, the species of the larger genera are related to each other in the same manner as the varieties of any one species are related to each other. No naturalist pretends that all the species of a genus are equally distinct from each other; they may generally be divided into sub-genera, or sections, or lesser groups. As Fries has well remarked, little groups of species are generally clustered like satellites around other species. And what are varieties but groups of forms, unequally related to each other, and clustered round certain forms—that is, round their parent-species? Undoubtedly there is one most important point of difference between varieties and species; namely, that the amount of difference between varieties, when compared with each other or with their parent-species, is much less than that between the species of the same genus. [p49]

Finally, varieties cannot be distinguished from species,—except, first, by the discovery of intermediate linking forms; and, secondly, by a certain indefinite amount of difference between them; for two forms, if differing very little, are generally ranked as varieties, notwithstanding that they cannot be closely connected; but the amount of difference considered necessary to give to any two forms the rank of species cannot be defined. In genera having more than the average number of species in any country, the species of these genera have more than the average number of varieties. In large genera the species are apt to be closely, but unequally, allied together, forming little clusters round other species. Species very closely allied to other species apparently have restricted ranges. In all these respects the species of large genera present a strong analogy with varieties. And we can clearly understand these analogies, if species once existed as varieties, and thus originated; whereas, these analogies are utterly inexplicable if species are independent creations. [p49f]

Again, it may be asked, how is it that varieties, which I have called incipient species, become ultimately converted into good and distinct species, which in most cases obviously differ from each other far more than do the varieties of the same species? How do those groups of species, which constitute what are called distinct genera, and which differ from each other more than do the species of the same genus, arise? All these results, as we shall more fully see in the next chapter, follow from the struggle for life. Owing to this struggle, variations, however slight and from whatever cause proceeding, if they be in any degree profitable to the individuals of a species, in their infinitely complex relations to other organic beings and to their physical conditions of life, will tend to the preservation of such individuals, and will generally be inherited by the offspring. The offspring, also, will thus have a better chance of surviving, for, of the many individuals of any species which are periodically born, but a small number can survive. I have called this principle, by which each slight variation, if useful, is preserved, by the term Natural Selection, in order to mark its relation to man's power of selection. [Chapter III, p51f]

Hence, as more individuals are produced than can possibly survive, there must in every case be a struggle for existence, either one individual with another of the same species, or with the individuals of distinct species, or with the physical conditions of life. [p53]

But the struggle will almost invariably be most severe between the individuals of the same species, for they frequent the same districts, require the same food, and are exposed to the same dangers. In the case of varieties of the same species, the struggle will generally be almost equally severe, and we sometimes see the contest soon decided... [p60]

In order that any great amount of modification should be effected in a species, a variety, when once formed, must again, perhaps after a long interval of time, vary or present individual differences of the same favourable nature as before; and these must be again preserved, and so onward, step by step. [p66]

It must have struck most naturalists as a strange anomaly that, both with animals and plants, some species of the same family and even of the same genus, though agreeing closely with each other in their whole organisation, are hermaphrodites, and some unisexual. But if, in fact, all hermaphrodites do occasionally intercross, the difference between them and unisexual species is, as far as function is concerned, very small. [p77f]

Intercrossing plays a very important part in nature by keeping the individuals of the same species, or of the same variety, true and uniform in character. It will obviously thus act far more efficiently with those animals which unite for each birth; but, as already stated, we have reason to believe that occasional intercrosses take place with all animals and plants. Even if these take place only at long intervals of time, the young thus produced will gain so much in vigour and fertility over the offspring from long-continued self-fertilisation, that they will have a better chance of surviving and propagating their kind; and thus in the long run the influence of crosses, even at rare intervals, will be great. With respect to organic beings extremely low in the scale, which do not propagate sexually, nor conjugate, and which cannot possibly intercross, uniformity of character can be retained by them under the same conditions of life, only through the principle of inheritance, and through natural selection which will destroy any individuals departing from the proper type. If the conditions of life change and the form undergoes modification, uniformity of character can be given to the modified offspring, solely by natural selection preserving similar favourable variations. [p79]

Isolation, also, is an important element in the modification of species through natural selection. In a confined or isolated area, if not very large, the organic and inorganic conditions of life will generally be almost uniform; so that natural selection will tend to modify all the varying individuals of the same species in the same manner. Intercrossing with the inhabitants of the surrounding districts will, also, be thus prevented. Moritz Wagner has lately published an interesting essay on this subject, and has shown that the service rendered by isolation in preventing crosses between newly-formed varieties is probably greater even than I supposed. But from reasons already assigned I can by no means agree with this naturalist, that migration and isolation are necessary elements for the formation of new species. The importance of isolation is likewise great in preventing, after any physical change in the conditions such as of climate, elevation of the land, &c., the immigration of better adapted organisms; and thus new places in the natural economy of the district will be left open to be filled up by the modification of the old inhabitants. Lastly, isolation will give time for a new variety to be improved at a slow rate; and this may sometimes be of much importance. If, however, an isolated area be very small, either from being surrounded by barriers, or from having very peculiar physical conditions, the total number of the inhabitants will be small; and this will retard the production of new species through natural selection, by decreasing the chances of favourable variations arising. [p79f]

Although isolation is of great importance in the production of new species, on the whole I am inclined to believe that largeness of area is still more important, especially for the production of species which shall prove capable of enduring for a long period, and of spreading widely. Throughout a great and open area, not only will there be a better chance of favourable variations, arising from the large number of individuals of the same species there supported, but the conditions of life are much more complex from the large number of already existing species; and if some of these many species become modified and improved, others will have to be improved in a corresponding degree, or they will be exterminated. Each new form, also, as soon as it has been much improved, will be able to spread over the open and continuous area, and will thus come into competition with many other forms. Moreover, great areas, though now continuous, will often, owing to former oscillations of level, have existed in a broken condition; so that the good effects of isolation will generally, to a certain extent, have concurred. Finally, I conclude that, although small isolated areas have been in some respects highly favourable for the production of new species, yet that the course of modification will generally have been more rapid on large areas; and what is more important, that the new forms produced on large areas, which already have been victorious over many competitors, will be those that will spread most widely, and will give rise to the greatest number of new varieties and species. They will thus play a more important part in the changing history of the organic world. [p80f]

I have been struck with the fact, that if any animal or plant in a state of nature be highly useful to man, or from any cause closely attracts his attention, varieties of it will almost universally be found recorded. These varieties, moreover, will often be ranked by some authors as species. Look at the common oak, how closely it has been studied; yet a German author makes more than a dozen species out of forms, which are almost universally considered by other botanists to be varieties; and in this country the highest botanical authorities and practical men can be quoted to show that the sessile and pedunculated oaks are either good and distinct species or mere varieties. [p88]

Hence modifications of structure, viewed by systematists as of high value, may be wholly due to the laws of variation and correlation, without being, as far as we can judge, of the slightest service to the species. [p110]

… on the view that species are only strongly marked and fixed varieties, we might expect often to find them still continuing to vary in those parts of their structure which have varied within a moderately recent period, and which have thus come to differ. Or to state the case in another manner: the points in which all the species of a genus resemble each other, and in which they differ from allied genera, are called generic characters; and these characters may be attributed to inheritance from a common progenitor, for it can rarely have happened that natural selection will have modified several distinct species, fitted to more or less widely different habits, in exactly the same manner:—and as these so-called generic characters have been inherited from before the period when the several species first branched off from their common progenitor, and subsequently have not varied or come to differ in any degree, or only in a slight degree, it is not probable that they should vary at the present day. On the other hand, the points in which species differ from other species of the same genus are called specific characters; and as these specific characters have varied and come to differ since the period when the species branched off from a common progenitor, it is probable that they should still often be in some degree variable,—at least more variable than those parts of the organisation which have for a very long period remained constant. [p115]

Divergence of Character.

The principle, which I have designated by this term, is of high importance, and explains, as I believe, several important facts. In the first place, varieties, even strongly-marked ones, though having somewhat of the character of species—as is shown by the hopeless doubts in many cases how to rank them—yet certainly differ from each other far less than do good and distinct species. Nevertheless, according to my view, varieties are species in the process of formation, or are, as I have called them, incipient species. How, then, does the lesser difference between varieties become augmented into the greater difference between species? That this does habitually happen, we must infer from most of the innumerable species throughout nature presenting well-marked differences; whereas varieties, the supposed prototypes and parents of future well-marked species, present slight and ill-defined differences. Mere chance, as we may call it, might cause one variety to differ in some character from its parents, and the offspring of this variety again to differ from its parent in the very same character and in a greater degree; but this alone would never account for so habitual and large a degree of difference as that between the species of the same genus. [p127]

The view commonly entertained by naturalists is that species, when intercrossed, have been specially endowed with sterility, in order to prevent their confusion. This view certainly seems at first highly probable, for species living together could hardly have been kept distinct had they been capable of freely crossing. The subject is in many ways important for us, more especially as the sterility of species when first crossed, and that of their hybrid offspring, cannot have been acquired, as I shall show, by the preservation of successive profitable degrees of sterility. It is an incidental result of differences in the reproductive systems of the parent-species. [p209]

The fertility of varieties, that is of the forms known or believed to be descended from common parents, when crossed, and likewise the fertility of their mongrel offspring, is, with reference to my theory, of equal importance with the sterility of species; for it seems to make a broad and clear distinction between varieties and species. [p209f]

It is certain, on the one hand, that the sterility of various species when crossed is so different in degree and graduates away so insensibly, and, on the other hand, that the fertility of pure species is so easily affected by various circumstances, that for all practical purposes it is most difficult to say where perfect fertility ends and sterility begins. I think no better evidence of this can be required than that the two most experienced observers who have ever lived, namely Kölreuter and Gärtner, arrived at diametrically opposite conclusions in regard to some of the very same forms. It is also most instructive to compare—but I have not space here to enter into details—the evidence advanced by our best botanists on the question whether certain doubtful forms should be ranked as species or varieties, with the evidence from fertility adduced by different hybridisers, or by the same observer from experiments made during different years. It can thus be shown that neither sterility nor fertility affords any certain distinction between species and varieties. The evidence from this source graduates away, and is doubtful in the same degree as is the evidence derived from other constitutional and structural differences. [p210f]

I believe that species come to be tolerably well-defined objects, and do not at any one period present an inextricable chaos of varying and intermediate links" [p213]

By the term systematic affinity is meant, the general resemblance between species in structure and constitution. Now the fertility of first crosses, and of the hybrids produced from them, is largely governed by their systematic affinity.

This is clearly shown by hybrids never having been raised between species ranked by systematists in distinct families; and on the other hand, by very closely allied species generally uniting with facility. But the correspondence between systematic affinity and the facility of crossing is by no means strict. A multitude of cases could be given of very closely allied species which will not unite, or only with extreme difficulty; and on the other hand of very distinct species which unite with the utmost facility. ...

No one has been able to point out what kind or what amount of difference, in any recognisable character, is sufficient to prevent two species crossing. [p215f]

It may be urged, as an overwhelming argument, that there must be some essential distinction between species and varieties, inasmuch as the latter, however much they may differ from each other in external appearance, cross with perfect facility, and yield perfectly fertile offspring. With some exceptions, presently to be given, I fully admit that this is the rule. But the subject is surrounded by difficulties, for, looking to varieties produced under nature, if two forms hitherto reputed to be varieties be found in any degree sterile together, they are at once ranked by most naturalists as species. [p226]

From these facts it can no longer be maintained that varieties when crossed are invariably quite fertile. From the great difficulty of ascertaining the infertility of varieties in a state of nature, for a supposed variety, if proved to be infertile in any degree, would almost universally be ranked as a species;—from man attending only to external characters in his domestic varieties, and from such varieties not having been exposed for very long periods to uniform conditions of life;—from these several considerations we may conclude that fertility does not constitute a fundamental distinction between varieties and species when crossed. The general sterility of crossed species may safely be looked at, not as a special acquirement or endowment, but as incidental on changes of an unknown nature in their sexual elements. [p229]

First crosses between forms, sufficiently distinct to be ranked as species, and their hybrids, are very generally, but not universally, sterile. The sterility is of all degrees, and is often so slight that the most careful experimentalists have arrived at diametrically opposite conclusions in ranking forms by this test. The sterility is innately variable in individuals of the same species, and is eminently susceptible to the action of favourable and unfavourable conditions. The degree of sterility does not strictly follow systematic affinity, but is governed by several curious and complex laws. It is generally different, and sometimes widely different in reciprocal crosses between the same two species. It is not always equal in degree in a first cross and in the hybrids produced from this cross.

In the same manner as in grafting trees, the capacity in one species or variety to take on another, is incidental on differences, generally of an unknown nature, in their vegetative systems, so in crossing, the greater or less facility of one species to unite with another is incidental on unknown differences in their reproductive systems. There is no more reason to think that species have been specially endowed with various degrees of sterility to prevent their crossing and blending in nature, than to think that trees have been specially endowed with various and somewhat analogous degrees of difficulty in being grafted together in order to prevent their inarching in our forests. [p233]

Naturalists, as we have seen, try to arrange the species, genera, and families in each class, on what is called the Natural System. But what is meant by this system? Some authors look at it merely as a scheme for arranging together those living objects which are most alike, and for separating those which are most unlike; or as an artificial method of enunciating, as briefly as possible, general propositions,—that is, by one sentence to give the characters common, for instance, to all mammals, by another those common to all carnivora, by another those common to the dog-genus, and then, by adding a single sentence, a full description is given of each kind of dog. The ingenuity and utility of this system are indisputable. But many naturalists think that something more is meant by the Natural System; they believe that it reveals the plan of the Creator; that unless it be specified whether order in time or space, or both, or what else is meant by the plan of the Creator, it seems to me that nothing is thus added to our knowledge. Expressions such as that famous one by Linnæus, which we often meet with in a more or less concealed form, namely, that the characters do not make the genus, but that the genus gives the characters, seem to imply that some deeper bond is included in our classifications than mere resemblance. I believe that this is the case, and that community of descent—the one known cause of close similarity in organic beings—is the bond, which though observed by various degrees of modification, is partially revealed to us by our classifications. [p319f]

The importance, for classification, of trifling characters, mainly depends on their being correlated with many other characters of more or less importance. The value indeed of an aggregate of characters is very evident in natural history. Hence, as has often been remarked, a species may depart from its allies in several characters, both of high physiological importance, and of almost universal prevalence, and yet leave us in no doubt where it should be ranked. Hence, also, it has been found that a classification founded on any single character, however important that may be, has always failed; for no part of the organisation is invariably constant. The importance of an aggregate of characters, even when none are important, alone explains the aphorism enunciated by Linnæus, namely, that the characters do not give the genus, but the genus gives the characters; for this seems founded on the appreciation of many trifling points of resemblance, to slight to be defined. [p321]

All the foregoing rules and aids and difficulties in classification may be explained, if I do not greatly deceive myself, on the view that the Natural System is founded on descent with modification;—that the characters which naturalists consider as showing true affinity between any two or more species, are those which have been inherited from a common parent, all true classification being genealogical;—that community of descent is the hidden bond which naturalists have been unconsciously seeking, and not some unknown plan of creation, or the enunciation of general propositions, and the mere putting together and separating objects more or less alike.

But I must explain my meaning more fully. I believe that the *arrangement* of the groups within each class, in due subordination and relation to each other, must be strictly genealogical in order to be natural; but that the *amount* of difference in the several branches or groups, though allied in the same degree in blood to their common progenitor, may differ greatly, being due to the different degrees of modification which they have undergone; and this is expressed by the forms being ranked under different genera, families, sections, or orders. [p323]

With species in a state of nature, every naturalist has in fact brought descent into his classification; for he includes in his lowest grade, that of species, the two sexes; and how enormously these sometimes differ in the most important characters, is known to every naturalist: scarcely a single fact can be predicated in common of the adult males and hermaphrodites of certain cirripedes, and yet no one dreams of separating them. ... The naturalist includes as one species the various larval stages of the same individual, however much they may differ from each other and from the adult, as well as the so-called alternate generations of Steenstrup, which can only in a technical sense be considered as the same individual. He includes monsters and varieties, not from their partial resemblance to the parent-form, but because they are descended from it.

As descent has universally been used in classing together the individuals of the same species, though the males and females and larvæ are sometimes extremely different; and as it has been used in classing varieties which have undergone a certain, and sometimes a considerable, amount of modification, may not this same element of descent have been unconsciously used in grouping species under genera, and genera under higher groups, all under the so-called natural system? I believe it has been unconsciously used; and thus only can I understand the several rules and guides which have been followed by our best systematists. As we have no written pedigrees, we are forced to trace community of descent by resemblances of any kind. Therefore we choose those characters which are the least likely to have been modified, in relation to the conditions of life to which each species has been recently exposed. Rudimentary structures on this view are as good as, or even sometimes better than, other parts of the organisation. We care not how trifling a character may be—let it be the mere inflection of the angle of the jaw, the manner in which an insect's wing is folded, whether the skin be covered by hair or feathers—if it prevail throughout many and different species, especially those having very different habits of life, it assumes high value; for we can account for its presence in so many forms with such different habits, only by inheritance from a common parent. We may err in this respect in regard to single points of structure, but when several characters, let them be ever so trifling, concur throughout a large group of beings having different habits, we may feel almost sure, on the theory of descent, that these characters have been inherited from a common ancestor; and we know that such aggregated characters have especial value in classification.

We can understand why a species or a group of species may depart from its allies, in several of its most important characteristics, and yet be safely classed with them. This may be safely done, and is often done, as along as a sufficient number of characters, let them be ever so unimportant, betray the hidden bond of community of descent. Let two forms have not a single character in common, yet, if these extreme forms are connected together by a chain of intermediate groups, we may at once infer their community of descent, and we put them all into the same class. As we find organs of high physiological importance—those which serve to preserve life under the most diverse conditions of existence—are generally the most constant, we attach especial value to them; but if these same organs, in another group or section of a group, are found to differ much, we at once value them less in our classification. We shall presently see why embryological characters are of such high classificatory importance. Geographical distribution may sometimes be brought usefully into play in classing large genera, because all the species of the same genus, inhabiting any distinct and isolated region, are in all probability descended from the same parents. [p325f]

When the views advanced by me in this volume, and by Mr. Wallace, or when analogous views on the origin of species are generally admitted, we can dimly foresee that there will be a considerable revolution in natural history. Systematists will be able to pursue their labours as at present; but they will not be incessantly haunted by the shadowy doubt whether this or that form be a true species. This, I feel sure and I speak after experience, will be no slight relief. The endless disputes whether or not some fifty species of British brambles are good species will cease. Systematists will have only to decide (not that this will be easy) whether any form be sufficiently constant and distinct from other forms, to be capable of definition; and if definable, whether the differences be sufficiently important to deserve a specific name. This latter point will become a far more essential consideration than it is at present; for differences, however slight, between any two forms, if not blended by intermediate gradations, are looked at by most naturalists as sufficient to raise both forms to the rank of species.

Hereafter we shall be compelled to acknowledge that the only distinction between species and well-marked varieties is, that the latter are known, or believed, to be connected at the present day by intermediate gradations whereas species were formerly thus connected. Hence, without rejecting the consideration of the present existence of intermediate gradations between any two forms, we shall be led to weigh more carefully and to value higher the actual amount of difference between them. It is quite possible that forms now generally acknowledged to be merely varieties may hereafter be thought worthy of specific names; and in this case scientific and common language will come into accordance. In short, we shall have to treat species in the same manner as those naturalists treat genera, who admit that genera are merely artificial combinations made for convenience. This may not be a cheering prospect; but we shall at least be freed from the vain search for the undiscovered and undiscoverable essence of the term species. [p371]

After the Origin

If we consider all the races of man as forming a single species, his range is enormous; but some separate races, as the Americans and Polynesians, have very wide ranges. It is a well-known law that widely-ranging species are much more variable than species with restricted ranges; and the variability of man may with more truth be compared with that of widely-ranging species, than with that of domesticated animals. [*Descent of Man*, chapter 2, (second edition, Darwin 1871)]

… how, it may be asked, have species arisen in a state of nature? The differences between natural varieties are slight; whereas the differences are considerable between the species of the same genus, and great between the species of distinct genera. How do these lesser differences become augmented into the greater difference? How do varieties, or as I have called them, incipient species, become converted into true and well-defined species? [*Variation of Plants and Animals under Domestication* (Darwin 1998), published in 1868 (Darwin 1868) and revised in 1875 (Darwin 1875, Vol I: 45)]

... that the sterility of distinct species when crossed, and of their hybrid progeny, depends exclusively on the nature of their sexual elements, and not on any differences in their structure or general constitution. ... That excellent observer, Gärtner, likewise concluded that species when crossed are sterile owing to differences confined to their reproductive systems [*Op. cit.* vol. II, p168f]

Passing over the fact that the amount of external difference between two species is no sure guide to the degree of their mutual sterility, so that similar differences in the case of varieties [of domestic animals and plants—*JSW*] would be no sure guide, we know that with species the cause lies exclusively in differences in their sexual constitution. [*Op. cit.* vol. II, p172]

The writer of the article in referring to my words "the preservation of useful variations of pre-existing instincts" adds "the question is, whence these variations?" Nothing is more to be desired in natural history than that some one would be able to answer such a query. But as far as our present subject is concerned, the writer probably will admit that a multitude of variations have arisen, for instance in colour and in the character of the hair, feathers, horns, &c., which are quite independent of habit and of use in previous generations. It seems far from wonderful, considering the complex conditions to which the whole organisation is exposed during the successive stages of its development from the germ, that every part should be liable to occasional modifications: the wonder indeed is that any two individuals of the same species are at all closely alike. [Short letter to *Nature* (Darwin 1873)]

I am surprised that Agassiz did not succeed in writing something better. How absurd that logical quibble—"if species do not exist, how can they vary?" As if any one doubted their temporary existence. How coolly he assumes that there is some clearly defined distinction between individual differences and varieties. It is no wonder that a man who calls identical forms, when found in two countries, distinct species, cannot find variation in nature. Again, how unreasonable to suppose that domestic varieties selected by man for his own fancy (p. 147) should resemble natural varieties or species. [Darwin to Asa Gray, August 11, 1860 (Darwin 1888a, Vol. 2: 333)]

How painfully (to me) true is your remark, that no one has hardly a right to examine the question of species who has not minutely described many. [Darwin to Hooker, September 1849 (Darwin 1888b: 39)]

Alfred Russel Wallace (1823–1913)

Wallace has been called the "forgotten naturalist" of evolution (Raby 2001; Shermer 2002; Wilson 2000). His paper on natural selection, although he didn't call it that and never liked the term, later recommending instead the equally ill-fated phrase of Spencer's "survival of the fittest" to Darwin, triggered Darwin to act after a 20 year delay in publishing his ideas. A clear and clever thinker, Wallace later took several of his premises to their logical conclusion and rejected the idea that selection could account for human cognitive and cultural capacities, and adopted Spiritualism.

Nothwithstanding his apparent heresy, Wallace's ideas about species were very influential. He adopted whole-heartedly the view that species were identifiable in terms of their not interbreeding or recombining into another population of forms.

> If there is no other character, that fact is one of the strongest arguments against the independent creation of species, for why should a special act of creation be required to call into existence an organism differing only in degree from another which has been produced by existing laws? If an amount of permanent difference, represented by any number up to 10, may be produced by the ordinary course of nature, it is surely most illogical to suppose, and very hard to believe, that an amount of difference represented by 11 required a special act to call it into existence. [(Wallace 1858; quoted in Kottler 1978: 294)]

> [A species is:] An assemblage of individuals which have become somewhat modified in structure, form, and constitution, so as to adapt them to slightly different conditions of life; which can be differentiated from allied assemblages; which reproduce their like; which usually breed together; and, perhaps, when crossed with their near allies, always produce offspring which are more or less sterile *inter se*. [*Darwinism* (Wallace 1889): 167, (quoted in Romanes 1895, vol 2: 236)]

> One of the best and most orthodox definitions is that of Pritchard [*sic*], the great ethnologist, who says, that "separate origin and distinctness of race, evinced by a constant transmission of some characteristic peculiarity of organization" constitutes a species. Now leaving out the question of "origin" which we cannot determine, and taking only the proof of separate origin, "the constant transmission of some characteristic peculiarity of organization," we have a definition which will compel us to neglect altogether the amount of difference between any two forms, and to consider only whether the differences that present themselves are permanent. The rule, therefore, I have endeavoured to adopt is, that when the difference between two forms inhabiting separate areas seems quite constant,

when it can be defined in words, and when it is not confined to a single peculiarity only, I have considered such forms to be species. [*Contributions to the Theory of Natural Selection* (Wallace 1870): 142]

Species are merely those strongly marked races or local forms which when in contact do not intermix, and when inhabiting distinct areas are generally believed to have had a separate origin, and to be incapable of producing a fertile hybrid offspring. … it will be evident that we have no means whatever of distinguishing so-called "true species" from the several modes of variation here pointed out, and into which they so often pass by an insensible gradation. [*Op. cit.* p161]

John Stevens Henslow (1796–1861)

One-time teacher and mentor to Darwin, Henslow remained a correspondent and confidant of Darwin's until his death.

If I understand him [Richard Owen], he thinks the "Becoming" of species (I suppose he means the *producing* of species) a somewhat rapid and not a slow process—but he seems to think them *progressive* organised [sic] out of previously organized beings—{analogous (?) to minerals (simple and compound) out of ± 60 Elements}. [Letter to Darwin, 5 May 1860 (Barlow 1967))

… how frequently Naturalists were at fault in regarding as *species*, forms which had (in some cases) been shown to be varieties, and how legitimately Darwin had deduced his *inferences* from positive experiment" [Letter 10 May 1860 to Hooker, which was then passed on to Darwin, *op. cit.*]

Asa Gray (1810–1888)

Gray was once a co-worker of Agassiz's, but he was among the first converts to the new evolutionary view in America, and was a regular correspondent of Darwin's. In fact, Darwin revealed his ideas to Gray well before publishing the *Origin*. Gray was a well-regarded botanist, and so his comments on species are a good indicator of the views in play at different times.

625. Species in biological natural history is a chain or series of organisms of which the links or component individuals are parent and offspring. Objectively, a species is the totality of beings which have come from one stock, in virtue of that most general fact that likeness is transmitted from parent to progeny. Among the many definitions, that of A. L. Jussieu is one of the briefest and best, since it expresses the fundamental conception of a species, *i. e.*, the perennial succession of similar individuals perpetuated by generation.

626. The two elements of species are: (1) community of origin; and, (2) similarity of the component individuals. But the degree of similarity is variable, and the fact of genetic relationship can seldom be established by observation or historical evidence. It is from the likeness that the naturalist ordinarily decides that such and such individuals belong to one species. Still the likeness is a consequence of the genetic relationship; so that the latter is the real foundation of species.

627. No two individuals are exactly alike; and offspring of the same stock may differ (or in their progeny may come to differ) strikingly in some particulars. So two or more forms which would have been regarded as wholly distinct are sometimes proved to be of one species by evidence of their common origin, or more commonly are inferred to be so from the observation of a series of intermediate forms which bridge over the differences. Only observation can inform us how much difference is compatible with a common origin. The general result of observation is that plants and animals breed true from generation to generation within certain somewhat indeterminate limits of variation; that those individuals which resemble each other within such limits interbreed freely, while those with wider differences do not. Hence, on the one hand, the naturalist recognizes *varieties* or differences within time species, and on the other *genera* and other superior associations, indicative of remoter relationship of the species themselves. (Gray 1879: 317f)

Pierre Trémaux (1818–1895)

Ethnologist, archaeologist and architect, Trémaux was influential to a degree on Marx, among others, and held that geography formed species. He is remembered mainly for introducing photography into field anthropology. His book quoted below is an overly ambitious attempt to explain all change as result of pressure and repulsion, similar in some ways to Lamarck. He held that life was itself a movent force. The section from which this is taken is an early review of definitions of species, mostly those in the French literature.

The word *species* is perhaps the most commonly encountered term in the study of the natural sciences. It is the first word and the last according to a celebrated zoologist[1], and on the day we are totally in control of it, we will be very close to the millennium of science in general.

...

The most formidable task of the naturalist, says Candolle, is to deal with species.

Species are fixed and of independent creation, says one school. They are variable and related among themselves, responds another. "We show one fact, fixity as far as can be traced," responds the first, "and you, you do nothing but expose an hypothesis without proof and one that encounters the strongest objections."

All schools agree in recognizing the greatest difficulties, and posing the most serious objections, to the transformations of beings in order to pass from one species to another.

"These difficulties are so grave," says Mr Darwin[2], "that I have been shaken by them for a long time... What geological research has been unable to reveal to us is the existence of numerous degrees of transition, as close to one another as recent varieties, and relating among themselves all known species. This is the most important of the objections that could be raised against my theory."

"This objection is decisive," cries Mr Flourens[3]. "This eternal distinction between species is both the greatest marvel and the greatest mystery in nature."

"The mystery of mysteries," Others have said before him.[4]

"The secret that God has reserved to himself," says Mr Duroy.[4]

Faced with such testimony one must acknowledge that the difficulty is great. But really, the solution is simple; so simple that, for not having found it, one could accuse our predecessors and all antiquity of blindness, if we did not keep in mind the view of the most eminent [scientists] of our era, which will help us towards this mystery of mysteries. Here is this point of view:

"The characters of the parent [species] could be *different*. In this case the corresponding characters of the child [species] would be a *resultant* [*résultante*], that is to say, in reality, a new character that did not exist in the father nor in the mother.

"The same cause operating in each generation would evidently produce effects of the same nature. Simple heredity, direct and immediate, is in certain respects a source of new variation of the first type."

Before going on, dear reader, let me ask a question: does this reasoning seem just, or does it seem false? In other words, the father, the mother, and the child, do they constitute a source of variation so that there are three types instead of two? ...

You may respond most probably, yes, and our predecessors, and all of antiquity, would be absolved.

But wait! A hair, an imperceptible thread, has slipped into this reasoning and vitiated it; the defect is this: the father and the mother belong to the current generation that is going to disappear, the child to the generation that succeeds

---

4   [John Herschel, in a letter quoted by Charles Babbage in the *Ninth Bridgewater Treatise*, (Babbage 1837) and quoted by Darwin in the opening paragraph of the *Origin. Transl.*]

it. Thus, the generation which disappears *unites* as a type in the following generation.

This is quite contrary to the point of view that we have just cited: instead of seeing a source of *variation* of types, we find a cause of their *unification*. And, if we have a large group of individuals, the definitive resultant will be the same after a certain number of generations.

Thus, let us take a highly variable population, where we find the most perfect and most imperfect types, the blackest as well as the whitest tints, in a word everything that is most disparate. At the moment of interbreeding, what will happen: if two similar organisms unite, their offspring will continue the same type; but if one of the progenitors is beautiful, and the other ugly, one black and the other white, the generation that will follow will be a *resultant*, which is to say, an intermediate type. And if, as is probable, different types cross only in part in each generation, a certain number of generations will be sufficient for all of the types to increasingly *unite* and in the end form only one average type, apart from causes of variation that arise from other sources.

Readers, permit me another question: now that this great fact is clear and you see that we are on the road to discovery, do you understand the mystery of the formation of species? ... perhaps not yet. So our predecessors can be totally excused for not having recognized it, even though all of modern science is superfluous for its recognition and serves only for its confirmation.

Well then, what is this great mystery?

Two principal notions have served to define species: resemblance between individuals and the ability to reproduce. To the extent that the first of these conditions, which is only a consequence of the second, can be abandoned in order to concentrate on the second, one approaches the solution to the problem.

For Laurent de Jussieu, the species is a succession of individuals that are entirely similar, perpetuated through reproduction.

For Buffon, the species is the constant succession of individuals which reproduce, and the character of the species is continual interfertility [*fécondité*].

Blainville defines species: the individual repeated in time and space.[5]

According to Lamarck, the species is a collection of similar individuals, which reproduction perpetuates in the same state so long as the circumstances of their situation does not change enough to cause variation in their habits, their character, and their form.

Of definitions of species, there are as many as there are naturalists, and this is the inevitable consequence of ignorance of the principle on which the species is

---

5   [Henri Marie Ducrotay de Blainville (1777–1850)]

based. Even so, Flourens contributes, with Buffon, Illiger, Koelreuter, Goertner and others, to help us take a step towards the question, by characterising a species only by *continuous fecundity*. At this point, one need only distinguish the effect from the cause.

We understand perfectly what happens when two organisms cross that are as different as possible. The product in general is an intermediate type, or about average between the two extremes.[5] Also, in the sum of many crossings, it is even more exactly the same. We also know that if the organisms are too different, one from the other, they are unable to produce properly similar descendants, or even to procreate at all. The effect of interfertility and the limits of its action, so to speak, two results of experience that are perfectly known, are all that is needed to form and reform species, whenever there occurs a cause of the modification of the established order.

With these facts, so simple, and known by all, the mystery reveals itself. God has delivered to us his secret.

First the definition of species, the essence of the great secret:

The species is composed of all the organisms which, able to procreate together, actually do so and group their descendents together in an intermediate type.

[1] I. Geoffroy Saint-Hilaire, *Histoire naturelle génerale des règnes organiques*, vol. II, p. 349. [(Geoffroy Saint Hilaire 1859)]

[2] *Op. cit.*[*Origin of species*] p241 and p422.

[3] Pages 41 and 98 of his book critique of the origin of species.

[4] *Histoire de la formation du sol français* (first part).

[5] "I give to the product of crossed unions, the name *Mongrel*, because *Mongrel* seems to me to comprise half of each of the two producing species" (Flourens, *op. cit.*, page 109).

Numerous observations have, indeed, noted that the types of the descendant are in general intermediate between those of the generators of different races or species, and if there are some rare exceptions, they could not have a marked influence on the group.

[*The Origin and Transformations of Man and Other Beings*
(Trémaux 1865: 127–136), translation by Gareth J. Nelson and John Wilkins]

... therefore, *the species is made up of all the beings which, being able to procreate together, in virtue of this fact group their descendants under an average type and the extent of possible and continuous fecundity, which precisely makes the variation which separates the closely related species!* [(Trémaux 1878: 48, italics original)[6]]

---

6 ... *donc, l'espèce est constituée par tous les êtres qui, pouvant procréer ensemble, groupent par ce fait leurs descendants sous un type moyen et l'étendue des fécondités possibles et continues, fait précisément l'écart qui sépare les espèces voisines!*...

Gregor Mendel o. s. a. (1822–1884)

Müller-Wille and Orel (2007) show how Mendel often used the term *Art* for "variety" and that he made no clear distinction between species and varieties in his 1866 paper.

> If we adopt the strictest definition of a species [*Artbegriffes*], according to which only those individuals belong to a species which under precisely the same circumstances display precisely similar characters, no two of these <u>varieties</u> could be referred to one species. According to the opinion of experts, however, the majority belong to the species Pisum sativum; while the rest are regarded and classed, some as sub-species of P. sativum, and some as independent species, such as P. quadratum, P. saccharatum, and P. umbellatum. The positions, however, which may be assigned to them in a classificatory system are quite immaterial for the purposes of the experiments in question. It has so far been found to be just as impossible to draw a sharp line between the hybrids of species [*Species*] and varieties [*Varietäten*] as between species and varieties themselves. [Mendel 1866: 10 (quoted in Müller-Wille and Orel 2007:174f)]

John Campbell, the Duke of Argyll (1845–1914)

Campbell's *The Reign of Law* was widely read and in part is a reaction to the new evolutionary view of the world. In it, he defends the view that Natural Law is insufficient to explain the world, and that the Supernatural—that is, God— must be introduced to complete explanations. However, while he is defending the Paleyesque view of design in the world, Campbell clearly takes Darwin seriously, and does not deny that evolution occurs, according to Natural Law. His comment here represents the sum of his criticism of Darwin's theory.

> It will be seen, then, that the principle of Natural Selection has no bearing whatever on the Origin of Species, but only on the preservation and distribution of species when they have arisen. I have already pointed out that Mr. Darwin does not always keep this distinction clearly in view... [(Argyll 1884: 240)]

August Weismann (1834–1914)

Weismann is an interesting thinker. His "Barrier", which is the notion for which he is best remembered, is that heredity is restricted to the germline cells (the reproductive gametes of sperm and egg) and does not involve any changes taken from the somatic (bodily) cells. He is taken to have shown that the neo-Lamarckian conception popular from around 1880 or so, in which influences to the body affect the nature of heredity (a view Darwin held to an extent with Use and Disuse), is inviable, and he did the famous experiment of docking generations of mouse tails to show that tails remained as strongly inherited as ever.

Often held to be a strict adaptationist, Weismann was one of those who Romanes sneeringly called "neo-Darwinians", or "ultra-Darwinians"—more Darwinian than Darwin. However, the first excerpt here indicates that as far as species were concerned, he held a view not dissimilar to the Founder Effect, that species are formed by random samplings of the original populational variation.

> … there are very variable species and very constant species, and it is obvious that colonies which are founded by a very variable species can hardly ever remain exactly identical with the ancestral species; and that several of them will turn out differently, even granting that the conditions of life be exactly the same, for no colony will contain all the variants of the species in the same proportion, but at most only a few of them, and the result of mingling these must ultimately result in the development of a somewhat different form in each colonial area. [*The evolution theory*, Vol II (Weismann 1904: 286)]

> All the individual members of these series are connected by intermediate forms in such a manner that a long period of constancy of forms seems to be succeeded by a shorter period of transformation, from which again a relatively constant form arises.

> We see, therefore, that the idea of species is fully justified in a certain sense; we find indeed at certain times a breaking up of the fixed specific type, the species becomes variable, but soon the medley of forms clears up again, and a new constant form arises—*a new species*, which remains the same for a long series of generations, until ultimately it too begins to waver, and is transformed once more But if we were to place side by side the cross-sections of this genealogical tree at different levels, we should only see several well-defined species between which no intermediate forms could be recognized; these would only be found in the intermediate strata. [*The evolution theory* (Weismann 1904, Vol. II: 305)]

> … if the step from one species to the next succeeding one does not depend on adaptation, then the greater steps to genera, families, and orders cannot be referred to it either, since these can only be thought of as depending upon a long-continued splitting up of species. [*Op. cit.*, p306]

> … the species is essentially a complex of adaptations, of modern adaptations which have been recently acquired, and of inherited adaptations handed down from long ago—a complex which might well have been other than it is, and

indeed must have been different if it had originated under the influence of other conditions of life. [*Op. cit.,* p307]

But of course species are not exclusively complicated systems of adaptations, for they are at the same time 'variation complexes,' the individual components of which are not all adaptive, since they do not all reach the limits of the useful or the injurious. [*Loc. cit.*]

[Selection in the "germ plasm" may] ... give rise to correlative variations in determinants next to them or related to them in any way, and that these may possess the same stability as the primary variation. This seems to me sufficient reason why biologically unimportant characters may become constant characters of the species. [*Loc. cit.*]

George John Romanes (1848–1894)

Romanes was Darwin's only direct student. Darwin is said to have remarked how wonderful it was that he was so young when they first met. He was considered the heir-apparent to Darwin. He maintained an "orthodox" interpretation of Darwin's theories, right down to pangenesis and sexual selection, which were either rejected or ignored by many later "Darwinians" like Wallace and Weismann, and to an extent Lankester. So far as I can tell, he is the originator of the [false, in my view] claim that the *Origin* does not give an account of the origin of species, so often repeated today. Here also is the beginnings of a list of "types of species concepts" that fed into the Essentialist Story.

When we remember the incalculable number of animal and vegetable species, living and extinct, we immediately feel the necessity for some much more general explanation of their existence than is furnished by supposing that their mutual sterility, which constitutes their most general or constant distinction, was in every case due to some incidental effect produced on the generative system by uniform conditions of life. To say nothing of the antecedent improbability, that in all these millions and millions of cases the reproductive system should happen to have been affected in this peculiar way by the merely negative condition of uniformity; there is, as it seems to me, the overwhelming consideration that, at the time when a variety is first forming, this condition of prolonged exposure to uniform conditions of life must necessarily be absent as regards that variety; yet this is just the time when we must suppose that the infertility with its parent form arose... [*Physiological Selection* (Romanes 1886)]

... it appears to me obvious that the theory of natural selection has been mis-named; it is not, strictly speaking, a theory of the origin of *species*: it is a theory of the origin—or rather of the cumulative development—of *adaptations*, whether these be morphological, physiological, or psychological, and whether they occur in species only, or likewise in genera, families, orders, and classes. [*Ibid*]

According to the general theory of evolution, which in this paper is taken for granted, the distinction between varieties and species is only a distinction of degree; and the distinction is mainly, as well as most generally, that of mutual sterility, whether absolute or partial. [*Ibid*]

According to this theory, species are but records of a sufficient degree of sterility having arisen with parent forms to admit of the varietal form not becoming swamped by intercrossing. Now, the degree of sterility required for this purpose would not be the same in all cases, seeing that in some cases other conditions might be present to assist in the prevention of intercrossing, as we shall see later on. [*Ibid*]

As now repeatedly observed, the theory of natural selection is not, properly speaking, a theory of the origin of species: it is a theory of the development of adaptive structures. Only if species always differed from one another in respect of adaptive structures would natural selection be a theory of the origin of species. But, as we have already seen, species do not always, or even generally, thus differ from one another. In what, then, do they differ?

They differ, first, chiefly and most generally, in respect of their reproductive systems; this, therefore, I will call the primary difference. Next, they differ in an endless variety of more or less minute details of structure, which are sometimes of an adaptive character, and sometimes not. These, therefore, I will call secondary differences. [*Ibid*]

Well, whether or not it is absurd and preposterous to consider climatic variations in connexion with the origin of species, will depend, and depend exclusively, on what it is that we are to understand by a species. Hitherto I have assumed, for the sake of argument, that we all know what is meant by a species. But the time has now come for showing that such is far from being the case. And as it would be clearly absurd and preposterous to conclude anything with regard to specific characters before agreeing upon what we mean by a character as specific, I will begin by giving all the logically possible definitions of a species.

1. A group of individuals descended by way of natural generation from an originally and specially created type.

This definition may be taken as virtually obsolete.

2. A group of individuals which, while fully fertile inter se, *are sterile with all other individuals—or, at any rate, do not generate fully fertile hybrids.*

This purely physiological definition is not nowadays [230] entertained by any naturalist. Even though the physiological distinction be allowed to count for something in otherwise doubtful cases, no systematist would constitute a species on such grounds alone. Therefore we need not concern ourselves with this definition, further than to observe that it is often taken as more or less supplementary to each of the following definitions.

3. A group of individuals which, however many characters they share with other individuals, agree in presenting one or more characters of a peculiar kind, with some certain degree of distinctness.

In this we have the definition which is practically followed by all naturalists at the present time. But, as we shall presently see more fully, it is an extremely lax definition. For it is impossible to determine, by any fixed and general rule, what degree of distinctness on the part of peculiar characters is to be taken as a uniform standard of specific separation. So long as naturalists believed in special creation, they could feel that by following this definition (3) they were at any rate doing their best to tabulate very real distinctions in nature—viz. between types as originally produced by a supernatural cause, and as subsequently more or less modified (i.e. within the limits imposed by the test of cross-fertility) by natural causes. But evolutionists are unable to hold any belief in such real distinctions, being confessedly aware that all distinctions between species and varieties are purely artificial. So to speak, they well know that it is they themselves who create species, by determining round what degrees of differentiation their diagnostic boundaries shall be drawn. And, seeing that these [231] degrees of differentiation so frequently shade into one another by indistinguishable stages (or, rather, that they *always* do so, unless intermediate varieties have perished), modern naturalists are well awake to the impossibility of securing any approach to a uniform standard of specific distinction. On this account many of them feel a pressing need for some firmer definition of a species than this one—which, in point of fact, scarcely deserves to be regarded as a definition at all, seeing that it does not formulate any definite criterion of specific distinctness, but leaves every man to follow his own standards of discrimination. Now, as far as I can see, there are only two definitions of a species which will yield to evolutionists the steady and uniform criterion required. These two definitions are as follows.

4. A group of individuals which, however many characters they share with other individuals, agree in presenting one or more characters of a peculiar and hereditary kind, with some certain degree of distinctness.

It will be observed that this definition is exactly the same as the last one, save in the addition of the words "and hereditary." But, it is needless to say, the addition of these words is of the highest importance, inasmuch as it supplies exactly that objective and rigid criterion of specific distinctness which the preceding definition lacks. It immediately gets rid of the otherwise hopeless wrangling over species as "good" and "bad," or "true" and "climatic," of which (as we have seen) Kerner's essay is such a remarkable outcome. Therefore evolutionists have [232] more and more grown to lay stress on the hereditary Character of such peculiarities as they select for diagnostic features of specific distinctness. Indeed it is not too much to say that, at the present time, evolutionists in general recognize this character as, theoretically, indispensable to the constitution of a species. But it is likewise not too much to say that, practically, no one of our systematic naturalists has hitherto concerned himself with this matter. At all events, I do not know of any who has ever taken the trouble to ascertain by experiment, with regard to any of the species which he has constituted, whether the peculiar characters on which his diagnoses have been founded are, or are not, hereditary. Doubtless the labour of constituting (or, still more, of *re*-constituting) species on such a basis of experimental inquiry would be insuperable; while, even if it could be accomplished, would prove undesirable, on account of the chaos it would produce in our specific nomenclature. But, all the same, we must remember that this nomenclature as we now have it—and, therefore, the partitioning of species as we have now made them—has no reference to the criterion of heredity. Our system of distinguishing between species and varieties is not based upon the definition which we are now considering, but upon that which we last considered—frequently coupled, to some undefinable extent, with No. 2.

5.   There is, however, yet another and closer definition, which may be suggested by the ultra-Darwinian school, who maintain the doctrine of natural selection as the only possible cause of the origin of species, namely:–

[233] *A group of individuals which, however many characters they share with other individuals, agree in presenting one or more characters of a peculiar, hereditary, and adaptive kind, with some certain degree of distinctness.*

Of course this definition rests upon the dogma of utility as a necessary attribute of characters *quâ* specific—i.e. the dogma against which the whole of the present discussion is directed. Therefore all I need say with reference to it is, that at any rate it cannot be adduced in any argument where the validity of its basal dogma is in question. For it would be a mere begging of this question to argue that every species must present at least one peculiar and adaptive character, because, according to definition, unless an organic type does present at least one such character, it is not a specific type. Moreover, and quite apart from this, it is to be hoped that naturalists as a body will never consent to base their diagnostic work on what at best must always be a highly speculative extension of the Darwinian theory. While, lastly, if they were to do so with any sort of consistency, the precise adaptation which each peculiar character subserves, and which because of this adaptation is constituted a character of specific distinction, would have to be determined by actual observation. For no criterion of specific distinction could be more vague and mischievous than this one, if it were to be applied on grounds of mere inference that such and such a character, because seemingly constant, must "necessarily" be either useful, vestigial, or correlated.

Such then, as far as I can see, are all the [234] definitions of a species that are logically possible[1]. Which of them is chosen by those who maintain the necessary usefulness of all specific characters? Observe, it is for those who maintain this doctrine to choose their definition: it is not for me to do so. My contention is, that the term does not admit of any definition sufficiently close and constant to serve as a basis for the doctrine in question—and this for the simple reason that species-makers have never agreed among themselves upon any criterion of specific distinction. My opponents, on the other hand, are clearly bound to take an opposite view, because, unless they suppose that there is some such definition of a species, they would be self-convicted of the absurdity of maintaining a great generalization on a confessedly untenable basis. For example, a few years ago I was allowed to raise a debate in the Biological Section of the British Association on the question to which the present chapters are devoted.

But the debate ended as I had anticipated that it must end. No one of the naturalists present could give even the vaguest definition of what was meant by [235] a species—or, consequently, of a character as specific. On this account the debate ended in as complete a destruction as was possible of the doctrine that all the distinctive characters of every species must necessarily be useful, vestigial, or correlated. For it became unquestionable that the same generalization admitted of being made, with the same degree of effect, touching all the distinctive characters of every "snark."

[1] It is almost needless to say that by a definition as "logical" is meant one which, while including all the differentiae of the thing defined, excludes any qualities which that thing may share in common with any other thing. But by definitions as "logically possible" I mean the number of separate definitions which admit of being correctly given of the same thing from different points of view. Thus, for instance, in the present case, since the above has been in type the late M. Quatrefages' posthumous work on *Darwin et ses Précurseurs François* has been published, and gives a long list of definitions of the term "species" which from time to time have been enunciated by as many naturalists of the highest standing as such (pp. 186-187). But while none of these twenty or more definitions is logical in the sense just defined, they all present one or other of the differentiae given by those in the text.

[*Darwin and After Darwin* (Romanes 1895, Vol. 2: 229–234)]

Edwin Ray Lankester (1847–1929)

Lankester was an influential British zoologist who carried on the Darwinian tradition, as well as publishing several popular science books. He was also a friend of Marx's and acted as a pallbearer at Marx's funeral.

> Linnaeus himself recognised the purely subjective character of his larger groups; for him species were, however, objective: "There are," he said, "just so many species as in the beginning the Infinite Being created." It was reserved for a philosophic zoologist of the nineteenth century (Agassiz, *Essay on Classification*, 1859) to maintain dogmatically that genus, order, and class were also objective facts capable of precise estimation and valuation. This climax was reached at the very moment when Darwin was publishing the *Origin of Species* (1859), by which universal opinion has been brought to the position that species, as well as genera, orders, and classes, are the subjective expressions of a vast ramifying pedigree in which the only objective existences are individuals, the apparent species as well as higher groups being marked out, not by any distributive law, but by the purely non-significant operation of human experience, which cannot transcend the results of death and decay. ["The History and Scope of Zoology", (ch. IX in Lankester 1890: 319)]
>
> [Lankester was] … inclined to think that we should discard the word species not merely momentarily but altogether. [(Poulton 1903: 62)]

Ernst Heinrich Philipp August Haeckel (1834–1919)

Haeckel is a controversial figure in the history of evolutionary biology. Although he called his views *Darwinismus*, he also asserted that Goethe and Lamarck were equal wellsprings for his thinking, and he was something of a late ideal morphologist. Haeckel invented phylogenetics, but was arbitrary and theory driven in his reconstructions of evolutionary trees. However, his terms and many of his ideas affected subsequent theory extensively, even though his "Law of Recapitulation" (*ontogeny recapitulates phylogeny*) was rejected almost immediately. His views on species seem to be fairly typical of the post-evolutionary period.

> Endless disputes arose among the "pure systematizers" on the empty question, whether the form called a species was "a good or bad species, a species or a variety, a sub-species or a group", without the question being even put as to what these terms really contained and comprised. If they had earnestly endeavoured to gain a clear conception of the terms, they would long ago have perceived that they have no absolute meaning, but are merely stages in the classification, or

systematic categories, and of relative importance only. [*The Evolution of Man* (Haeckel 1883 vol 1: 115)]

Thomas Henry Huxley (1825 –1895)

Huxley, like Haeckel, was more influenced by the *Naturphilosophen* or ideal morphologists than by ordinary taxonomic issues, and in some ways is more like his opponent Owen than Darwin. He seems to have had no problem with the use of morphological criteria for classification, but he does at least indicate there might be a physiological cause for species-rank.

> Animals and plants are divided into groups, which become gradually smaller, beginning with a KINGDOM, which is divided into SUB-KINGDOMS; then come the smaller divisions called PROVINCES; and so on from a PROVINCE to a CLASS, from a CLASS to an ORDER, from ORDERS to FAMILIES, and from these to GENERA, until at length we come to the smallest groups of animals which can be defined one from the other by constant characters, which are not sexual; and these are what naturalists call SPECIES in practice, whatever they may do in theory.

> If in a state of nature you find any two groups of living beings, which are separated one from the other by some constantly-recurring characteristic, I don't care how slight or trivial, so long as it is defined and constant, and does not depend on sexual peculiarities, then all naturalists agree in calling them two species; that is what is meant by the word species—that is to say, it is, for the practical naturalist, a mere question of structural differences.[1]

> ...
> [1] I lay stress here on the *practical* signification of "Species." Whether a physiological test between species exist or not, it is hardly ever applicable by the practical naturalist.

> [(Huxley 1906: 226f)]

> Living beings, whether animals or plants, are divisible into multitudes of distinctly definable kinds, which are morphological species. They are also divisible into groups of individuals, which breed freely together, tending to reproduce their like, and are physiological species. [Anonymous review of the *Origin*, 1859 (Huxley 1893: 50)]

> [Naturalists employ the term *species* in a double sense to denote "two very different orders of relations".] When we call a group of animals, or of plants, a species, we may imply thereby either, that all these animals and plants have some common peculiarity of form or structure; or, we may mean that they possess some common functional character. That part of biological science which deals with form and structure is called Morphology—that which concerns itself with function, Physiology—so that we may conveniently speak of these two senses

or aspects of "species"—the one as morphological, the other as physiological. ["Darwin on The Origin of Species" published in the *Westminster Review* of 1860 (Forsdyke 2001: 31) p302]

Richard F. Clarke, S. J. (1839–1900)

Writer of a Catholic textbook of logic, which indicates that logical species are being here identified with natural species, perhaps formally for the first time. This is published at the height of the Neo-Scholastic movement begun in 1874.

SPECIES contains the *whole essence* of the individual, and a concept which thus includes the whole essence is said to be a *species* in reference to each and all of the individuals contained under the general term. [*Logic* (Clarke 1895: 171f)]

Essences are indivisible, say scholastic logicians, as well as *immutable*. They cannot be changed, and we cannot think of them as changed, without an anomaly presenting itself in the nature, an element of which has been thus reversed. [*Op. cit.* 183]

Edward Bagnell Poulton (1856–1943)

Although the term "species problem" was introduced by Bateson, Poulton's 1903 essay "What is a species?" marks the beginning of the debate of the present day. He introduced a number of terms, only one of which—*sympatry*—survives to the present debate.

1. Forms having certain structural characters in common distinguishing them from the forms of other groups. Groups thus defined by the Linnaean method of *Diagnosis* may be conveniently called *Syndiagnostic* (σύν, together; διάγνωσις, distinction).

2. Forms which freely interbreed together. These may be conveniently called *Syngamic* (σύν, together; γάμος, marriage). Free interbreeding under natural conditions may be termed *Syngamy*; its cessation or absence, *Asyngamy* (equivalent to the Amixia of Weismann). [3]

3. Forms which have been shown by human observation to be descended from common ancestors or from a common parthenogenetic or self-fertilizing ancestor. Such groups may be called *Synepigonic* (σύν, together; ἐπίγαμος, descendant). Breeding from common parents or from a common parthenogenetic or self-fertilizing parent may be spoken of as *Epigony* or the production of *Epigonic* evidence.

4. Finally there is geographical distribution, of the utmost importance in the modification and origin of species and sub-species. Forms found together in certain geographical areas may be called *Sympatric* (σύν, together; πάτρια, native country). The occurrence of forms together may be termed *Sympatry*, and the discontinuous distribution of forms Asympatry.

My friend, Professor E. Ray Lankester, to whom I owe so much, in this as in many other subjects, is inclined to think that we should discard the word species not merely momentarily but altogether. Modern zoology having abandoned Linnaeus's conception of 'species' should, he considers, abandon the use of the word. In his opinion the 'origin' of species was really the abolition of species, and zoologists should now be content to describe, name, draw, and catalogue *forms*. Furthermore, the various groups of forms briefly defined above should be separately and distinctly treated by the zoologist, without confusion or inference from one to the other. The systematist should say, ' I describe and name certain forms *a*, *b*, & *c*'; and then he or another may write a separate chapter, as it were: 'I now show that the forms *ab*, *ac*, *ad* (form names) are syngamic': at another time he may give reason for regarding any of them as related by Epigony.

I fear that this suggestion is a 'counsel of perfection', impossible of attainment, although there would be many and great advantages in thus making a fresh start and in the abandonment of 'species', or rather the restriction of the word to the only meaning it originally possessed before it was borrowed from logic to become a technical term in zoology. If the main contention of this Address be accepted, that a species is a syngamic and synepigonic group of individuals, an objective reality however difficult to establish in practice, we have an additional powerful reason for the permanent use of the word.

…

Diagnosis, it will be maintained, is founded upon the conception that there is unbroken transition in the characters of the component individuals of a species. Underlying this idea are the more fundamental conceptions of species as groups of individuals related by Syngamy and Epigony.

In immense numbers of cases it will be shown that the component individuals of a species do not form an unbroken series, but one that is sharply broken at one or more points. At each of these breaks the older systematist made a new species, which the modern systematist has rejected, because in his day the more fundamental criteria have been actually established or by strong indirect evidence have been inferred. When the test of Diagnosis necessarily fails as it will be shown to do in many large groups of examples the appeal is made to Syngamy and Epigony.

Syngamy and Epigony are but two sides of the same phenomenon Reproduction. Although occasional union between individuals of distinct species may occur in nature, sometimes leading to the production of hybrid offspring, this is not the 'free interbreeding under natural conditions' which I have called Syngamy. Syngamy, thus defined, implies the production of normal offspring capable of continuing the species implies Epigony. As a practical criterion, the evidences of Syngamy are generally much easier to collect than those of Epigony. Both Syngamy and Epigony can be established by indirect evidence based on a sufficient number of accurate observations upon the habits and modes of occurrence of individuals. The criterion of Syngamy of course fails in the case of parthenogenetic and self-fertilizing species. In such cases, ... we are compelled to fall back on Epigony. This latter criterion may lead, although only in rare and exceptional instances, to an erroneous inference, when hybrids are mistaken for normal offspring.

There is a great advantage in the admission that Diagnosis is only a provisional criterion, inasmuch as the systematist would then continually seek and continually suggest the search for the more fundamental tests.

...

The conclusions set forth above, if hereafter established, lead to a belief in the reality of species. Unlike and apart from genera, families and other groups employed in our 'little systems' of classification which 'have their day and cease to be', not only do individuals stand out as objective realities, but equally real, though far less evident, are the societies into which individuals are bound together in space and time by Syngamy and Epigony.

This idea of a species is clearly expressed by Sir William Thiselton-Dyer, when he speaks of the older writers who employed 'the word species as a designation for the totality of individuals differing from all others by marks or characters which experience showed to be reasonably constant and trustworthy, as is the practice of modern naturalists'.

This conception of a species is founded upon transition. Whenever a set of individuals can be arranged, according to the characters fixed upon by the systematist, in a series without marked breaks, that set is regarded as a species. The two ends of the series may differ immensely, may diverge far more widely than the series itself does from other series; but the gradual transition proclaims it a single species. If transitions were all equally perfect, of course there would be no difficulty. But transitions are infinite in their variety; while the subjective element is obviously dominant in the selection of gaps just wide enough to constitute interspecific breaks, just narrow enough to fuse the species separated by some other writer, dominant also in the choice of the specific characters themselves.

...

The idea of a species as an interbreeding community, as syngamic, is, I believe, the more or less acknowledged foundation of the importance given to transition. ["What is a Species?" in *Essays on Evolution* (Poulton 1903: 60–68)]

Erich Wasmann S.J. (1859–1931)

Wasmann, a noted entomologist, represents the reaction to Haeckel's monism and anti-Catholicism. His text was part of an ongoing campaign he ran against Haeckel (Lustig 2002; Richards 2005), in the course of which he developed a view similar to Buffon and Cuvier, in which a "natural species" was very much larger than a Linnaean species. He accepted evolution within these limited natural species. Note in particular his distinction, as if it were widely accepted, between species of the physical kind and species of the logical kind. It was, in fact, widely accepted.

Linnaeus, who is regarded as the originator of our present conception of systematic species, and who, therefore, has been called the father of the theory of permanence, enunciated the following dictum: *Tot species numeramus, quot diversae formae in principio sunt creatae*—we reckon so many (systematic) species as there were different forms created in the beginning.

How must this dictum be worded to make it agree with the theory of evolution? According to it, systematic species of the present time do not represent the originally created forms, but are the result of a process of evolution, uniting the species of the present and the past in natural series of forms, the members of which are related to one another, and each of which points back to an original primitive form, whence it is derived.. If we designate each of these independent series of forms, not related to other series or families, as a natural species,[1] we can still assent to Linnaeus's dictum : *Tot species numeramus, quot diversae formae in principio sunt creatae*. We reckon so many natural species as there were primitive forms created in the beginning.[2] Each of these natural species has in the course of evolution differentiated itself into more or less systematic species. How many systematic species, genera, and families belong to a natural species, cannot yet be stated with certainty in most cases. Still less are we able to say how many natural species there are, i.e., how many lines of ancestry independent of one another. We must leave the decision to the phylogenetic research of future ages, if indeed it ever succeeds at arriving at one.

The varying degrees of capacity of evolution possessed by the primitive forms of the different natural species depend primarily upon the interior laws of evolution impressed upon their organic constitution; we are probably justified in regarding the chromatin substance of the germ-cells as the material designed to transmit these laws. ... The interaction of these interior factors in evolution and of the surrounding exterior influences, through which many kinds of adaptation came about, have produced the ramifications from the parent stock of the natural species, and they have been affected also by cross-breeding (amphimixis) and natural selection.

[1] A similar view regarding natural species has already been expressed by Father T. Pesch in his *Philosophia naturalis*, II, p. 334, in order to explain the facts supporting the theory of evolution. He quotes a number of passages from St. Thomas Aquinas and from Suarez in favour of his view. Of course, we are here speaking of the *species physicae* of natural philosophy, not of the *species metaphysicae* of logic. ... [A] mistake was made by Friese ... and Schroeder ..., who believe my distinction between systematic and natural species to be identical with that between biological and morphological species; the biological and the morphological species are but two different aspects of the systematic species, whilst the natural species comprises all the members of the same line of ancestry or pedigree, and is therefore much wider from the point of view of natural science. ...

[2] For readers who have studied philosophy, it is perhaps needless to remark again (as I do for the benefit of some of my critics), that the creation of the first organisms is not to be understood as a *creatio e nihilo*, but as a production of organisms out of matter. ...

[*Modern Biology and the Theory of Evolution*, German edition 1906, (Wasmann 1910: 296f)]

Is the fixity of the organic species, that prevails at the present time, to lead us to conclude that species are absolutely invariable, and that therefore no evolution can have taken place in their case? Such a conclusion would be premature, for, granted that an evolution took place in previous ages, the results of it might be exactly what we see about us in the Alluvial epoch in which we live. ... Palaeontology teaches plainly enough that, in previous ages also, comparatively long periods of fixity have alternated with shorter periods of transformation of organic forms. ...

If we are at the present moment living in a period of comparative fixity of organic forms, we may seek in vain for actual changes in the species around us; but that circumstance proves nothing against the theory of evolution. [*Ibid*, p311]

Karl Jordan (1861–1959)

Jordan's view of species relied on there being "bridgeless gaps" between them, and that they had internal causes for being maintains separate. Lotsy (below) later called these units "Jordanons".

Individuals connected by blood relationship form a single faunistic unit in an area ... the units, of which the fauna of an area is composed, are separated from each other by gaps which at this point are not bridged by anything.[7] [(Jordan 1905): 157]

The criteria of the *species* [he uses the Latin word "species" here—*JSW*] (= *Art*) concept are thus threefold, and each individual point is an applicable test: a species requires known traits, it does not beget descendents equally well with individuals of other species, and it does not coalesce into other species.[8] [*Loc. cit.*]

The principal criterion of the conception "species" is that species can exist together without fusing, no other barrier keeping them apart than their own organization. [(Rothschild and Jordan 1906; quoted in Mallet 1995: 296)]

William Bateson (1861–1926)

Bateson, before he became a noted Mendelian, set up what we now call the "species problem"—that is, of defining species under the theory of transmutation of species.

No definition of a Specific Difference has been found, perhaps because these Differences are indefinite and hence not capable of definition. But the forms of living things, taken at a moment, do nevertheless most certainly form a discontinuous series and not a continuous series. ...

The existence, then, of Specific Differences is one of the characteristics of the forms of living things. This is no merely subjective conception, but an objective, tangible fact. This is the first part of the problem.

In the next place, not only do Specific forms exist in Nature, but they exist in such a way as to fit the place in Nature in which they are placed; that is to say, the Specific form which an organism has, is *adapted* to the position which it fills. This again is a relative truth, for the adaptation is not absolute. [(Bateson 1894): 2f]

All constructive theories of evolution have been built upon the understanding that we know if the relation of varieties to species justifies the assumption that

---

7   Jordan (1905: 157): "Solche blutsverwandte Individuen bilden eine faunistische Einheit in einem Gebeit ..." which Jordan follows with the elided comment "zu welcher Einheit wir erfährungsgemäß alle andern Individuen des Gebeits rechnen müssen, welche ihnen gleichen", meaning roughly that we assign all individuals to these units when we empirically classify them as identical. Again, there is an epistemological element here that is overlooked. The subsequent statement Mayr quotes comes after a number of examples of this classification.

8   "Das Kriterium des Begriffs Species (=Art) ist daher ein dreifaches, und jeder einzelne Punkt ist der Prüfung zugänglich: Eine Art bat gewisse Körpermerkmale, erzeugt keine den Individuen anderer Arten gleich Nachkommen und verschmilzt nicht mit andern Arten." Thanks to Ian Musgrave for help with the translation.

the one phenomenon is a *phase* of the other, and that each species arises ... from another species either by one, or several, genetic steps. ... [However,] complete fertility of the results of intercrossing [between members of different "species"] is, and I think must rightly be regarded, as *inconsistent* with actual specific difference." [*Problems of Genetics* 1913 (quoted in Forsdyke 2001: 31)]

Hugo de Vries (1848–1935)

De Vries, like Bateson, struggled to define species in a Mendelian context. Like others he rejected the reality of Linnaean species, but instead proposed a notion—"elementary species", which are pure bred genetic groups that do not interbreed with other groups.

Species is a word, which has always had a double meaning. One is the systematic species, which is the unit of our system. But these units are by no means indivisible. ... These minor entities are called varieties in systematic works. ... Some of these varieties are in reality just as good as species, and have been "elevated," as it is called, by some writers, to this rank. This conception of the elementary species would be quite justifiable, and would get rid of all difficulties, were it not for one practical obstacle. The number of species in all genera would be doubled and tripled, and as these numbers are already cumbersome in many cases, the distinction of the native species of any given country would lose most of its charm and interest.

In order to meet this difficulty we must recognize two sorts of species. The systematic species are the practical units of the systematists and florists, and all friends of wild nature should do their utmost to preserve them as Linnaeus has proposed them. These units, however, are not really existing entities; they have as little claim to be regarded as such as genera and families. The real units are the elementary species; their limits often apparently overlap and can only in rare cases be determined on the sole ground of field-observations. Pedigree-culture is the method required and any form which remains constant and distinct from its allies in the garden is to be considered as an elementary species. ["Species and Varieties", 1904 (De Vries 1912: 11)]

Elementary Species in Nature.

What are species? Species are considered as the true units of nature by the vast majority of biologists. They have gained this high rank in our estimation principally through the influence of Linnaeus. They have supplanted the genera which were the accepted units before Linnaeus. They are now to be replaced, in their turn, by smaller types, for reasons which do not rest upon comparative studies but upon direct experimental evidence.

Biological studies and practical interests alike make new demands upon systematic botany. Species are not only the subject-material of herbaria and collections, but they are living entities, and their life-history and life-conditions command a gradually increasing interest. One phase of the question is to determine the easiest manner to deal with the collected forms of a country, and another feature is the problem as to what groups are real units and will remain constant and unchanged through all the years of our observations. [*Op. cit.* p32f]

Elementary species differ from their nearest allies by progressive changes, that is by the acquisition of some new character. The derivative species has one unit more than the parent. [*Op. cit.* p253]

Charles B. Bessey (1845–1915)

A leading American botanist, Bessey was a student of Asa Gray's, and the author of one of the first phylogenetic classifications of plants. He made these remarks at a 1908 symposium on species, in which the consensus seemed to be that species did not exist for evolutionary reasons.

As long as species were supposed to be actual things, "created as separate kinds at the beginning," that botanists "discovered," as explorers discover islands in the ocean, there was no serious "species question." A botanist might make a mistake, and announce the discovery of a new species, when he had merely found a variety of an old species; as an explorer might mistakenly announce the discovery of a new island, when as a matter of fact he had only seen an unfamiliar coast of a long-known island.

Nature produces individuals, and nothing more. She produces them in such countless numbers that we are compelled to sort them into kinds in order that we may be able to carry them in our minds. This sorting is classification—taxonomy. But right here we are in danger of misunderstanding the matter. We do not actually sort out our individuals. We imagine them sorted out. It is only to a very slight extent that the systematic botanist ever actually sorts out individuals. When he has a considerable number of individual dried plants in his herbarium, he may sort them out, but these are but an infinitesimal portion of all the individuals in the world that we imagine to be sorted, but that are actually unsorted.

So species have no actual existence in nature. They are mental concepts, and nothing more. They are conceived in order to save ourselves the labor of thinking in terms of individuals, and they must be so framed that they do save us labor. If they do not, they fail of their purpose. Here we perceive one of the principles which must control in the limitation of species. If we multiply our species unduly we approach too near to the individuals, and we are as much burdened as though we had not invented species. On the other hand, if we make too few species those we do make include so many variations from the type that confusion results again. We must steer a course between these two extremes. [(Bessey 1908: 218–219)]

Horace William Brindley Joseph (1867–1943)

H.W.B. Joseph's *Introduction to Logic* was crucial to the development of the Essentialism Story, as outlined by Winsor (2004). Although like Whately he was careful to distinguish between logical species with essences and natural species, the founders of the Essentialism Story did not make that distinction. Joseph was a well known Idealist philosopher as well as logician.

The difficulty of determining what attributes are essential to a substance, and therefore of discriminating between essence and property, does not however arise entirely from the seeming disconnexion of the attributes of a kind. It arises also, in the case at least of the organic, from the great variation to which a species is liable in divers individuals. Extreme instances of such variation are sometimes known as border varieties, or border specimens; and these border varieties give great trouble to naturalists, when they endeavour to arrange all individuals in a number of mutually exclusive species. For a long time the doctrine of the fixity of species, supported as well by the authority of Aristotle and of Genesis, as by the lack of evidence for any other theory, encouraged men to hope that there was a stable character common to all members of a species, and untouched by variation; and the strangest deviations from the type, excluded under the title of monstrosities or unnatural births, were not allowed to disturb the symmetry of the theory. Moreover, a working test by which to determine whether individuals were of different species, was furnished, as is well known, by the fertility of offspring; it being assumed that a cross between different species would always be infertile, as in the case of the mule, and that when a cross was uniformly infertile, the species were different. But now that the theory of organic evolution has reduced the distinction between varietal and specific difference to one of degree, the task of settling what is the essence of a species becomes theoretically impossible. ...

If [biological] species were fixed: if there were in each a certain nucleus of characters, that must belong to the members of any species either not at all or all in all: if it were only upon condition of exhibiting at least such a specific nucleus of characters that the functions of life could go on in the individual at all; then this nucleus would form the essence of the kind. But such is not the case. The conformity of an individual to the type of a particular species depends on the fulfilment of an infinity of conditions, and implies the exhibition of an infinity of correlated peculiarities, structural and functional, many of which, so far as we can see ... have no connexion one with another. There may be deviation from the type, to a greater or less degree, in endless directions; and we cannot fix by any hard and fast rule the amount of deviation consistent with being of the species... Hence for definition, such as we have it in geometry, we must substitute classification ... A classification attempts to establish types... [*Introduction to Logic*, 2nd edition (Joseph 1916: 81f, 88f)]

Johannes Paulus Lotsy (1867–1931)

A botanist of note, Lotsy's book *Evolution by means of hybridization* in 1916 caused a major stir. It was an attempt to use Mendelian genetics to identify species as groups formed by genetic identity. He defined taxonomic replacement terms for "species" which gained currency in the lead-up to the Synthesis, but were later treated as curiosities, such as Linneon and Jordanon.

All theories of evolution have, until quite recently, been guided by a *vague* knowledge of what a species is, and consequently have been vague themselves. [(Lotsy 1916: 14)]

Jordan *consequently discarded morphological comparison as a criterium for specific purity* and, falling back to Ray (whom he may or may not have known) *substituted for it: nulla certior ... quam distincta propagatio ex semine.* [*Op. cit.*, p21, italics original]

A species consists of the total of individuals of identical constitution unable to form more than one kind of gametes. [*Op. cit.*, p23, italics original]

Specific purity is indicated by the uniformity and identity of the F_1 generations obtained by crossing the individuals to be tested, RECIPROCALLY. [*Op. cit.*, p24, italics original]

LINNEON: *to replace the term species in the Linnaean sense, and to designate a group of individuals which resemble one another more than they do any other individuals.*

To establish a Linneon consequently requires careful morphological comparison only.

JORDANON: *to replace the term species in the Jordanian sense, viz: mikrospecies* [sic], *elementary species etc. and to designate a group of externally alike individuals which all propagate their kind faithfully, under conditions excluding contamination by crossing with individuals belonging to other groups, as far as these external characters are concerned, with the only exception of noninheritable modifications of these characters, caused by the influences of the surroundings in the widest sense, to which these individuals or those composing the progeny may be exposed.*

To establish a Jordanon, morphological comparison alone consequently does not suffice; the transmittability of the characters by which the form was distinguished, must be experimental breeders.

SPECIES: *to designate a group of individuals of identical constitution, unable to form more than one kind of gametes; all monogametic individuals of identical constitution consequently belonging to one species.* [*Op. cit.*, p27, italics original]

Göte Wilhelm Turesson (1892–1970)

Turesson's work on the norms of reaction of different plant species in different conditions started a major research program. As a result, he defined species in terms of their ecological forms and places as well as their morphology and descent.

a)   *Ecospecies*: An amphimict-population the constituents of which in nature produce vital and fertile descendents with each other giving rise to less vital or more or less sterile descendants in nature, however, when crossed with constituents of any other population. ...

b)   *Agamospecies*: An apomict-population the constituents of which, for morphological, cytological or other reasons, are to be considered as having a common origin. ...

c)   *Coenospecies*: A population-complex the constituents of which group themselves in nature in species units of lower magnitude on account of vitality and sterility limits having all, however, a common origin so far as morphological, cytological or experimental facts indicate such and origin. ...

... existence in nature of units of these different orders also makes it a logical impossibility to reach one standard definition of the "species". [(Turesson 1929: 332–333)]

Charles Tate Regan (1878–1943)

Regan, referred to by Dobzhansky as an "affable taxonomist" (Dobzhansky 1937: 311), was an icthyologist and museum curator at the British Museum (Natural History) where he was in charge of zoology and later became director. His reworking of the comment Darwin heard from Phillips has been elevated into a "position" about the nature of species, which it probably is not.

... 'community' is the right name for a number of similar individuals that live together and breed together. ["Organic Evolution", 1925 (Regan 1926; quoted in Trewavas 1973: 92)]

A species is a community, or a number of communities, whose distinctive morphological characters are, in the opinion of a competent systematist, sufficiently definite to entitle it, or them, to a specific name. [*Loc. cit.* (Regan 1926: 75), cited in Julian Huxley (1942: 157; see also Ghiselin 1997: 118; Trewavas 1973)]

In practice it often happens that geographic forms, representing each other in different areas, are given only subspecific rank, and that closely related forms, not easily distinguished, are given specific rank when they inhabit the same area but keep apart. [(quoted in Trewavas 1973: 92)]

Thomas Hunt Morgan (1866–1945)

Morgan, like many geneticists, held that species were abstractions and not real.

We should always keep in mind the fact that the individual is the only reality with which we have to deal, and that the arrangement of these into species, genera, families, etc. is only a scheme invented by man for purposes of classification. Thus, there is no such thing in nature as a species, except as a concept of forms more or less alike. [(Morgan 1903; quoted in Allen 1980: 359f)]

Arthur Paul Jacot (1890–1939?)

An entomologist specialising in mites, at Shantung Christian University in China, Jacot's paper was greatly influential on the ideas of Sewall Wright, whose shifting balance theory of speciation depended in part on the distinctions between species not being adaptive (Provine 1986: 290). "Habit" in this paper appears to mean "mode of life".

There seems to be no valid species, based on a single character. So-called physiological "species," differentiated by a single physiological character, are typical varieties, not subspecies, which are geographical units ... . Any one species in a genus is differentiated from every other species by a *set* of morphological, as well also as physiological and habitudinal characters. Therefore a species originates, not by a change of one character, but by several contemporaneous changes. [(Jacot 1932: 346)]

Habitat or any habit may be as much a specific character as the structural character, originating as an independent unit character, accompanying the morphological characters and subject to all the laws of specific unit characters in transmission and heredity ... . Because we can easily see, measure and describe the structural characters, we use them for species differentiation; the habits are lost in alcohol. It is the habits, however, that determine the numbers of a species in its food relations, protective relations, propagative relations. [*Ibid* p350]

Certainly it is not the structural, taxonomic ear-marks of the species that determine an animal's (or plant's) degree of adaptation to its environment or give it supremacy but its habits, more probably breeding habits. Natural (congeneric) selection is operative far more through habits (and habitat preference) than through superficial structure, and it is not until the morphological differences become so marked as to be of generic or family value (as reiterated by Robson) that they become of adaptational (selective) value and thus supersede the value of habits in this respect. Let us henceforth turn our eyes more and more, then, on habit studies as of *early* evolutionary significance, even though we must change from systematist to ecologist (sociologist). [*Ibid* p351]

On the above basis, a species may be described as life forms related to each other, by the possession of the same combination of structural characters which are usually so poorly developed as to be of no immediate adaptive or selective value in the process of evolution, as well as by the possession of a combination of similar physiological and habitudinal characters. The structural characters are so small or poorly developed that the environment has no effect on them. These various specific characters continue being modified (from within) and recombined (within the genus) to form, through time, more and more species until the generic area becomes so crowded as to bring about congeneric relations. [*Ibid* p352f]

To summarize: A sedentary or gregarious "parental" form (phylogenic genotype) throws off various combinations of characters, each at a different center, each of which, by the growth of reproduction, advances from its center like an eddy across the country, influenced in its spread chiefly by "barriers." Each offshoot begins from the parent at a different time and place and is held constant by heredity. The eddies of range extension may be in any direction or have a general trend, depending on local factors governing distribution of species. At any time, however, some of these radiating species may in the course of migration meet a confrère with which it is capable of interbreeding and thus in turn become the center of origin of a radically new type as a subgenus or genus. [*Ibid* p353f]

Ronald Aylmer Fisher (1890–1962)

More than anyone else, Fisher began the synthesis between Mendelian genetics and Darwinian theory with his book *The Genetical Theory of Natural Selection*, in which he reconciled the apparent discreteness of genes and mutations with the gradual variation required for natural selection to operate on. In the course of a brilliant but flawed work (the flaw being largely his acceptance of eugenics in the latter half of the book) he made some oft-overlooked comments on species that have proved to be remarkably prescient of the modern debate.

The contrast between sexual and asexual reproduction

A group of organisms in which sexual reproduction was entirely unknown might none the less evolve under the action of natural selection. This condition cannot, I believe, be ascribed with certainty to any known group. Yet, since it is impossible to draw any sharp distinction within a whole series of asexual processes, from individual growth at the one extreme, through the regeneration of injured or lost parts, to vegetative reproduction by budding; it is tempting to believe that asexual reproduction was the primitive condition of living matter, and that the sexual reproduction of the predominant types of organisms is a development of some special value to the organisms which employ it. In such an asexual group, systematic classification would not be impossible, for groups of related forms would exist which had arisen by divergence from a common ancestor. Species, properly speaking, we could scarcely expect to find, for each individual genotype would have an equal right to be regarded as specifically distinct, and no natural groups would exist bound together like species by a constant interchange of their germ-plasm.

The groups most nearly corresponding to species would be those adapted to fill so similar a place in nature that any one individual could replace another, or more explicitly that an evolutionary improvement in any one individual threatens the existence of the descendants of all the others. Within such a group the increase in numbers of the more favoured types would be balanced by the continual extinction of lines less fitted to survive, so that, just as, looking backward, we could trace the ancestry of the whole group back to a single individual progenitor, so, looking forward at any stage, we can foresee the time when the whole group then living will be the descendants of one particular individual of the existing population. [*The Genetical Theory of Natural Selection*, 1930 (Fisher 1930: 121)]

The nature of species

From genetic studies in the higher organisms it may be inferred, that whereas genetic diversity may exist, perhaps in hundreds of different loci, yet in the great majority of loci the normal condition is one of genetic uniformity. Unless this were so the concept of the wild type gene would be an indefinite one. ... In many loci the whole of the existing genes in the species must be the lineal descendants of a single favourable mutation.

The intimate manner in which the whole body of individuals of a single species are bound together by sexual reproduction has been lost sight of by some writers. Apart from the intervention of geographical barriers so recently that the races separated are not yet regarded as specifically distinct, the ancestry of each single individual, if carried back only for a hundred generations, must embrace practically all of the earlier period who have contributed appreciably to the ancestry of the present population. If we carry the survey back for 200, 1,000, or 10,000 generations, which are relatively short periods in the history of most species, it is evident that the community of ancestry must be even more complete. The genetical identity in the majority of loci, which underlies the genetic variability presented by most species, seems to supply the systematist with the true basis of his concepts of specific identity or diversity. In his *Contributions to the Study of Variation*, W. Bateson frequently hints at an argument, which evidently influenced him profoundly, to the effect that the discontinuity to be observed between different species must have owed its origin to discontinuities occurring in the evolution of each. His argument, so far as it can be traced from a work, which owed its influence to the acuteness less of its reasoning than of its sarcasm, would seem to be correct for purely asexual organisms, for in these it is possible to regard each individual, and not merely each specific type, as the last member of a series, the continuity or discontinuity of which might be judged by the differences which occur between parent and offspring; and so to argue that these provide an explanation of the diversity of distinct strains. In sexual organisms this argument breaks down, for each individual is not the final member of a single series, but of converging lines of descent which ramify comparatively rapidly throughout the entire specific group. The variations which exist within a species are like the differences in colour between different threads which have crossed and recrossed each other a thousand times in the weaving a single uniform fabric. [*Ibid* 123-124]

Fission of species

The close genetic ties which bind species together into single bodies bring into relief the problem of their fission a problem which involves complexities akin to those that arise in the discussion of the fission of the heavenly bodies, for the attempt to trace the course of events through intermediate states of instability, seems to require in both cases a more detailed knowledge than does the study of stable states. In many cases without doubt the establishment of complete or almost complete geographical isolation has at once settled the line of fission; the two separated moieties thereafter evolving as separate species, in almost complete independence, in somewhat different habitats, until such time as the morphological differences between them entitle them to 'specific rank'. It would, however, be contrary to the weightiest opinions to postulate that specific differentiation had always been brought about by geographic isolation almost complete in degree. In many cases it may safely be asserted that no geographic isolation at all can be postulated, although this view should not be taken as asserting that the habitat of any species is so uniformly favourable, both to the maintenance of population, and to migration, that no 'lines of weakness' exist, which, if fission is in any case imminent, will determine the most probable geographic lines of division. It is, of course, characteristic of unstable states that minimal causes can at such times produce disproportionate effects; in discussing the possibility of the fission of species without geographic isolation, it will therefore be sufficient if we can give a clear idea of the nature of the causes which condition genetic instability. [*Ibid*, 124–125]

John Arthur Thompson (1861–1933)

Thompson's book was a popular introduction for the lay-biologist, and stands as an example of the general view just prior to the Synthesis.

> To sum up: *A species is a group of similar individuals differing from other groups in a number of more or less true-breeding characters, greater than those which often occur within the limits of a family, and not the direct result of environmental or other nurtural influences. The members of a species are fertile with one another, but not readily with other species.* [*Biology for Everyman* (Thompson 1934 vol. 2: 1333f), italics original]

Theodosius Hryhorovych Dobzhansky (1900–1975)

Dobzhansky began the present debate with his essay "A critique of the species concept in biology" in 1935, in a philosophy journal, no less. He repeated his discussion in his famous, and groundbreaking, book in 1937, *Genetics and the Origin of Species*. After Fisher, his is the first book to begin the Synthesis in a general way. His view is the foundation for the so-called "biological" conception of species, as it is a reproductive isolation conception, but he added a "vertical" or time dimension, and restricted species to a local horizon.

... a species is a group of individuals fully fertile inter se, but barred from interbreeding with other similar groups by its physiological properties (producing either incompatibility of parents, or sterility of the hybrids, or both). [(Dobzhansky 1935: 353; cf. also Dobzhansky 1941: 312)]

Considered dynamically, the species represents that stage of evolutionary divergence, at which the once actually or potentially interbreeding array of forms becomes segregated into two or more separate arrays which are physiologically incapable of interbreeding. [(Dobzhansky 1935: 354)]

[Species formation] ... through a slow process of accumulation of genetic changes of the type of gene mutations and chromosomal reconstructions. This premise being granted, it follows that instances must be found in nature when two or more races have become so distinct as to approach, but not to attain completely, the species rank. The decision of a systematist in such instances can not but be an arbitrary one. [*Genetics and the Origin of Species* (Dobzhansky 1937: 310f)]

We find aggregations of numerous more or less clearly distinct biotypes, each of which is constant and reproduces its like if allowed to breed. These constant biotypes are sometimes called elementary species, but they are not united into integrated groups that are known as species in the cross-fertilizing [i.e., sexual— *JSW*] forms. The term "elementary species" is therefore misleading and should be discarded.

Which one of these ranks is ascribed to a given cluster is, however, decided by considerations of convenience, and the decision is in this sense purely arbitrary. In other words, the species as a category which is more fixed, and therefore less arbitrary than the rest, is lacking in asexual and obligatorily self-fertilizing organisms. ...

The binomial system of nomenclature, which is applied universally to all living beings, has forced systematists to describe "species" in the sexual as well as in the asexual organisms. Two centuries have rooted this habit so firmly that any thorough reform will meet with determined opposition. Nevertheless, systematists have come to the conclusion that sexual species and "asexual species" must be distinguished ... . In the opinion of the writer, all that is saved by this method is the word "species". A realization of the fundamental difference between the two kinds of "species" can make the species concept methodologically more valuable than it has been. [(Dobzhansky 1941: 320f)]

The classical race concept [of human races—*JSW*] was typological. ...

Typology is at the bottom of the vulgar notion that any so-called Negro in the United States ... has a basic and unremediable Negroid nature, just as any Jew partakes of some Jewishness, etc. There are no Platonic types of Negroidness or Jewishness or of every race of squirrel or butterflies. Individuals are not mere reflections of their racial types; individual differences are the fundamental biological realities. [*Op. cit.* p268f]

Hermann Joseph Muller (1890–1967)

A leading founder of modern genetics, Muller also denied the reality of species, as Morgan had and Haldane later did.

It becomes, then a matter of definition and of convenience, in any given series of cases, just where we decide to draw the line above which two groups will be distinguished as separate species, and below which they are denoted subspecies or races, since in nature there is no abrupt transition here. [(Muller 1940: 253)]

Although all the work agrees in showing that speciation represents no absolute stage in evolution, but is gradually arrived at, and intergrades imperceptibly into racial differentiation beneath it and generic differentiation above, nevertheless the concept [*of "species"—JSW*] represents a real stage, corresponding to something of significance in nature, in so far as we may identify it with the stage at which effective intercrossing stops. For divergence goes on very differently, and much more freely, between those which can and do cross, and it is therefore justifiable and useful, even though difficult, to make the species distinction, if it is made in such a way as to correspond so far as possible with this stage of separation. At the same time it must be recognized that the species are in flux, and that an adequate understanding of their relationships can be arrived at only on the basis of an understanding of the relationships between the minor groups and even between the individuals, supplemented by the study of the differences found through observations on the systematics of the larger groups. [*Ibid* p258]

Cyril Dean Darlington (1903–1981)

Darlington is in the tradition established by Morgan, Muller and to an extent Bateson, that species are not realities. This will be repeated by Haldane later (below).

There are many kinds of species and many kinds of discontinuities between species. Those differences depend on the different kinds of genetic systems at work in plants and animals; but they cannot be arranged in a simple table because they occur at different levels of integration. The uniformity of systematic nomenclature so convenient for description suggests to us that the species is a proper subject for inductive consideration. We feel we ought to have a 'species concept'. In fact there can be no species concept based on the species of descriptive convenience that will not ensnare its own author so soon as he steps outside the group from which he made the concept. The only valid principles are those that we can derive, not from fixed classes but from changing processes. To do this we must go beyond the species to find out what it is made of. We must proceed (by collaboration) to examine its chromosomal structure and system of reproduction in relation to its range of variation and ecological character. From them we can determine what is the genetic species of Ray, the unit of reproduction, a unit which cannot be used for summary diagnosis, but which can be used for discovering and relating the processes of variation and the principles of evolution. [(Darlington 1940: 158f)]

John Burdon Sanderson Haldane (1892–1964)

With Haldane the Mendelian tradition of denying that species are more than conveniences is well established.

I object to the term "species concept", which I think is misleading. … A species in my opinion is a name given to a group of organisms for convenience, and indeed of necessity. [J. B. S. Haldane (1956: 95)]

Julian Sorell Huxley (1887–1975)

Huxley, whose text *The Modern Synthesis* gave the movement a name and rallying point, was nevertheless critical of the so-called biological conception of species as reproductively isolated, noting the exceptions in plants. He settles on a pluralistic view, in which several conceptions are sufficient by themselves, but not necessary.

The dynamic point of view is an improvement, as is the substitution of incapacity to exchange genes for the narrower criterion of infertility: but even so, this definition cannot hold, for it still employs the lack of interbreeding as its sole criterion. "Interbreeding without appreciable loss of fertility" would apply to the great majority of animals, but not to numerous plants. In plants there are many cases of very distinct forms hybridizing quite competently even in the field. To deny many of these forms specific rank just because they can interbreed is to force nature into a human definition, instead of adjusting your definition to the facts of nature. Such forms are often markedly distinct morphologically and do maintain themselves as discontinuous groups in nature. If they are not to be called species, then species in plants must be deemed to differ from species in animals in every characteristic save sterility… [*The Modern Synthesis* (Huxley 1942: 162f)]

In general, it is becoming clear that we must use a combination of several criteria in defining species. Some of these are of limiting nature. For instance, infertility between groups of obviously distinct mean type is a proof that they are distinct species, although once more the converse is not true.

First, then, we have the problem involved in the origin of species. As a preliminary to that, logic demands that we should define the term. It may be that logic is wrong, and that it would be better to leave it undefined, accepting the fact that all biologists have a pragmatic idea at the back of their heads. It may even be that the word is undefinable. However, an attempt at a definition will be of service in throwing light on the difficulties of the biological as well of the logical problems involved. [*Op. cit.* p 154]

Thus in most cases a group can be distinguished as a species on the basis of the following points jointly: (i) a geographical area consonant with a single origin; (ii) a certain degree of constant morphological and presumably genetic difference from related groups; (iii) absence of intergradation with related groups. ... Our third criterion above, if translated from the terminology of the museum to that of the field, may thus be formulated as a certain degree of biological isolation from related groups. [*Op. cit.* p164f]

Thus we must not expect too much of the term species. In the first place, we must not expect a hard-and-fast definition, for since most evolution is a gradual process, borderline cases must occur. And in the second place, we must not expect a single or a simple basis for definition, since species arise in many different ways. [*Op. cit.* p167]

John Scott Lennox Gilmour (1906–1986)

Gilmour, Curator of the University Herbarium and Botanical Museum in Cambridge, was one of the first to introduce the logical positivist viewpoint to biology, and in particular to biological classification. Gilmour argued that classification had to be "operationalised", and that the term "species" was too fixed in creationist terms to be useful. In conjunction with J. Heslop Harrison, he proposed to replace all existing taxonomic terms with the term "deme", which was unfortunately co-opted by the Synthesis to mean a population of breeding organisms, where he and Harrison had meant for it to be a rank-neutral term, that depended on "clipping" sense-data in various atheoretical ways (Winsor 2000).

Now the classification of animals and plants, though based on different data, is essentially similar in principle to the classification of inanimate objects. That is to say, it consists in clipping together the mass of sense-data collectively classified as 'living things' into a logically coherent pattern for the purpose of making inductive generalizations concerning those data. The primary 'clip' for living sense-data is the concept of the individual. As is well known ... this concept breaks down in a number of cases, but for general purposes, and especially in the higher animals and plants, the individual can be taken as a convenient unit of classification. It should never be forgotten, however, that the individual is a concept, a rational construction from sense-data, and that the latter are the real objective material of classification. ["Taxonomy and Philosophy" (Gilmour 1940: 465f)]

The concept of a unit character, however, is a notoriously vague and relative one, and it would seem that taxonomic categories based on resemblance in the sum total of attributes can never be susceptible of precise definition. As a definition of species, then, I would suggest something on the following lines, and analogous definitions could be constructed for other categories. 'A species is a group of individuals which, in the sum of their attributes, resemble each other to a degree usually accepted as specific, the exact degree being ultimately determined by the more or less arbitrary judgment of taxonomists.'

Admittedly this definition, based as it is on resemblance in total attributes, is a very vague one, but any attempt to define a species more precisely in terms of particular attributes breaks down. It is true, of course, that certain types of attribute are particularly important at the specific level of differentiation, for instance interfertility and chromosome number, but it has proved impossible to use them as a basis for a generally accepted definition. [*Ibid*, p468f]

With regard to experimental taxonomy, there is no doubt that the term 'species', however consciously one tries to forget its past associations and to adapt it to modern ideas, still comes to us trailing clouds of special-creation glory, which, as Camp has indicated, tend to bedevil its use in an evolutionary setting. It is largely these wisps from the past that have caused so much controversy over the relationship between nomenclatural and experimental taxonomy. The problem of whether to adapt or to abandon a particular term is a familiar one in the history of thought, and which course should be taken depends on the adaptability of the term and the extent of the change that has taken place. ... On balance, the abandonment of the term 'species' as a category to express the findings of experimental taxonomy, reflecting a clean break with the non-evolutionary past, would seem to be the right policy. The reluctance of geneticists and others to take this step is, perhaps, bound up with a lingering belief that species are, in some way, 'real entities', not 'artificial constructs of the human mind' as are other taxa. This belief makes it unthinkable to abandon the term for such an important branch of biology as the study of micro-evolution. If, however, 'real entities' and 'artificial constructs are regarded, not as mutually exclusive, but as alternative descriptions, each valid in its own context, this difficulty disappears. [(Gilmour 1958)].

❧ Section 4. Reproductive Isolation Conceptions

Let us now review the broad species conceptions presently in play. These are classes of concepts, here divided into *Reproductive Isolation Species Conceptions*, *Evolutionary Species Conceptions*, *Phylogenetic Species Conceptions*, *Ecological Species Conceptions*, and a trashcan categorical of *Other Species Conceptions* (Cracraft 1997; Mayden 1997; Wheeler and Meier 2000; Hey 2001; Mayden 2002).

We may arbitrarily mark the beginnings of the modern debate with Dobzhansky's classic 1935 essay, "A critique of the species concept in biology" (Dobzhansky 1935), although Poulton's essay is also crucial as is the later paper and book by Richards and Robson (Richards and Robson 1926; Robson 1928). It's not really so arbitrary—from Dobzhansky's essay and the book that followed it (Dobzhansky 1937) flowed both the present debate over what species are, and the birth of the Modern Synthesis. Dobzhansky's work was an attempt to take Darwin seriously about species by a working systematist not imbued with the British geneticists' deflationary tradition. And he introduced two of the major innovations in the debate—the idea of species as evolutionary players, and the idea that genetic exchange marked out these players. It seems that when typostrophic views of evolution were abandoned, the issue from the Great Chain of what divisions nature forced upon us and what divisions were of our own convenience came back to the fore. Later, Mayr described the Synthesis as being a "shared species problem" (Mayr and Provine 1980). It remains a shared problem.

The notion that species are kinds of organisms delineated not by the decisions of the classifiers but by the reproductive behaviors and results of the organisms themselves is an old one. As we have seen, it was suggested as part of John Ray's definition in 1688, and also by Buffon in 1748, but it goes back in the form of the generative conception of species to the time of the Greeks. To a greater or lesser degree it has been a component of nearly all species concepts since Linnaeus (his sexual system implicitly required reproductive isolation), and even now, it is a key component (Cracraft 2000) of phylogenetic species concepts and most other operational definitions.

The generative conception of species has been applied to living beings effectively back to Epicurus and the neo-Platonists. That is to say, there has *always* been a requirement not only of constancy of form, but of the reproduction of form, in definitions of living species. This is surprising, since the implication or tacit assumption of many discussions, such as Mayr's (1982) or Hull's (1965; 1988), is that species had been for much of the history of the concept arid definitional constructs based on "essences" (the "Essentialism Story"). At least since Locke there has always been a distinction between the *real* essence of a species (that is, what causes a species to *be* a species, its "real constitution") and the *nominal* essence (that is, how we know the species, describe it, define it, and apply a name to it).[1] It is not clear that "essence" in these cases plays only a definitional role—the Real Essence is not, by definition, definable.

Modern views of species are widely known and discussed, so we shall be brief in covering them, except for the ways that the most well-known, the reproductive isolation conceptions, developed. These conceptions often go by the name the "biological species concept". It is an unfortunate name, as most species concepts that are applied in biology are biological, and it doesn't really address the contrast Mayr, who so named it, wanted to make with "morphological species concepts", which were neither definitions of the concept of "species", nor more than identification criteria for individual species.

It is standard practice to distinguish these conceptions as if they were all unrelated, but clearly they form a family of conceptions, all based around reproductive isolation, by various mechanisms. So I term them in full *Reproductive Isolation Species Conceptions*, or RISCs for short.

Lack of interbreeding has played a role in many, if not most, conceptions of natural or biological species since the Greeks. It particularly comes to the fore when breeding practices are applied to species definitions, beginning, as we saw, in the late middle ages, but particularly in the period of the 17th century onwards. Mayr tended to overplay the novelty of his definition, and to deprecate those who came before him. Similarly, those who followed him either adhered

---

1   Hull has noted (in correspondence) that it is equally ahistorical and whiggist in turn for today's historians to apply current standards to the essentialism story historians of the 1960s; he, Mayr and Cain had what scholarly resources were then available. In large part due to their work, later research has identified the "missing links", and later work will no doubt overturn some of this and other modern work too. It should not be thought that this is a criticism of him or Mayr, etc., for failing to take into account later scholarship.

to his definition or a minor variation of it, or to themselves stress their novelty from his view.

Ernst Walter Mayr (1904–2005)

Mayr was a German ornithologist who had left Germany well before the war, and came to America (Hull 1988: 66f) and to the American Museum of Natural History and then to Harvard, after spending a number of years in New Guinea and the Solomon Islands studying bird populations and distributions. He was motivated to address the "species problem" because of the publication of another book, opposed to Dobzhansky's approach, by Richard Goldschmidt (1940), who became Mayr's *bête noire* for many years to come. Goldschmidt proposed that species evolved in a single step, through macromutations involving chromosomal repatterning to form "hopeful monsters" (pp390–393; it should be clear now that the term "monster" here refers to a sport or sudden variation from the type) in what is often called "saltation" (i.e., the opposite of *natura non facit saltum* quoted by Darwin). Goldschmidt repeatedly referred to species being separated by "bridgeless gaps" (c.f., p143), a term he took from Turesson (1922: 100), ignoring the fact that Turesson then went on to give a Darwinian account of species formation. Goldschmidt rejected the Darwinian idea that subspecific races were incipient species entirely, which seems to have motivated Mayr's ire and to have informed his allopatric account of speciation later.

Mayr was invited to give a series of talks on speciation as part of the Jesup Lectures in 1941 at Columbia University's Zoology Department. He was later invited to publish sufficient material to fill an entire volume for Columbia University Press after the other lecturer, Edgar Anderson, fell ill (p xvii, of the new Introduction to the 1999 reissue of his 1942). The result was the single most widely referred to monograph of the synthesis.

Basically this work is a discussion at length of the modes of speciation according to the best knowledge of the day, and much of what Mayr discussed remains valid. In the case of "ring-species" for example, his discussions remain the canonical ones (pp 180–185).[2] He also introduced several terms, including *allopatry* (p149), and *sibling species* (p151) which have worn well. But for our purposes,

---

2   However, some of the canonical examples are being disputed. Recently, molecular analysis of the *Larus argentatus* (Herring Gull) complex has indicated that they are, in fact, isolated gene pools (Liebers, Knijff, and Helbig 2004) and the *Parus major* (Great Tit) complex, while it does interbreed to some extent, is a good set of phylogenetic species (Kvist et al. 2003).

the most important aspect of the book is that here Mayr popularized the definition of species he had already given earlier (Mayr 1940).

A ... difficulty which confronts us in our attempt at a species definition is that there is, in nature, a great diversity of different kinds of species. Even if we do not consider such aberrant phenomena as the apomictic species in plants and the strains of bacteria, there is, even among animals, a great variety of different taxonomic situations which are generally classified as species. The question as to whether the species of birds, of corals, or protozoa, and of intestinal worms are the same kind of evolutionary phenomena is entirely justified. ... It may not be exaggeration if I say that there are probably as many species concepts as there are thinking systematists and students of speciation. [(Mayr 1942: 114–115)]

A species consists of a group of population which replace each other geographically or ecologically and of which the neighboring ones intergrade or interbreed wherever they are in contact or which are potentially capable of doing so (with one or more of the populations) in those cases where contact is prevented by geographical or ecological barriers.

Or shorter: Species are groups of actually or potentially interbreeding natural populations, which are reproductively isolated from other such groups. [*Systematics and the Origin of Species*, 1942 (Mayr 1940, 1942: 120)]

[Of Dobzhansky's definition] This is an excellent description of the process of speciation, but not a species definition. A species is not a stage of a process, but the result of a process. [*Op. cit.* p119]

The application of a biological species definition is possible only in well-studied taxonomic groups, since it is based on a rather exact knowledge of geographical distribution and on the certainty of the absence of interbreeding with other similar species. [*Op. cit.* p121]

Darwin thought of individuals when he talked of competition, struggle for existence among variants, and survival of the fittest in a particular environment. Such a struggle among individuals leads to a gradual change of populations, but not to the origin of new groups. It is now being realized that species originate in general through the evolution of entire populations. If one believes in speciation through individuals, one is by necessity an adherent of sympatric speciation, the two concepts being very closely connected. However, fewer and fewer situations are interpreted as evidence for sympatric speciation, as it is realized more and more clearly that reproductive isolation is required to make the gap between two incipient species permanent and that such reproductive isolation can develop only under exceptional circumstances between individuals of a single interbreeding population. [*Op. cit.* p190]

Noninterbreeding between populations is manifested by a gap. It is this gap between populations that coexist (are sympatric) at a single location at a given time that delimits the species recognized by a naturalist. Whether one studies birds, mammals, butterflies, or snails near one's home town, one finds species clearly delimited and sharply separated from all other species. This demarcation is sometimes referred to as the species delimitation in a non-dimensional system (a system without the dimensions of space and time). [(Mayr 1970: 14f, italics original)]

[A taxon is] ... the concrete object of classification. Any such group of populations is called a taxon if it is considered sufficiently distinct to be worthy of being formally assigned to a definite category in the hierarchical classification. A taxon is a taxonomic group of any rank that is sufficiently distinct to be worthy of being assigned to a definite category. [*Op. cit.* p14, italics original]

... species are reproductive communities. The individuals of a species of animals recognize each other as potential mates and seek each other for the purpose of reproduction. A multitude of devices insure intraspecific reproduction in all organisms ... . The species is also an ecological unit that, regardless of the individuals composing it, interacts as a unit with other species with which it shares the environment. The species, finally, is a genetic unit consisting of a large, intercommunicating gene pool, whereas the individual is merely a temporary vessel holding a small portion of the contents of the gene pool for a short time. [(Mayr 1963: 21)]

Species are groups of interbreeding natural populations that are reproductively isolated from other such groups. [(Quoted in Mayr 1970: 12, italics in original)]

[The biological species definition is "biological"] ... not because it deals with biological taxa, but because the definition is biological. It utilizes criteria that are meaningless as far as the inanimate world is concerned.

When difficulties are encountered, it is important to focus on the basic biological meaning of the species: A species is a protected gene pool. It is a Mendelian population that has its own devices (called isolating mechanisms) to protect it from harmful gene flow from other gene pools. Genes of the same gene pool form harmonious combinations because they become coadapted by natural selection. Mixing the genes of two different species leads to a high frequency of disharmonious gene combinations; mechanisms that prevent this are therefore favored by selection. [(Mayr 1970: 13)]

A species is a reproductive community of populations (reproductively isolated from others) that occupies a specific niche in nature. [(Mayr 1982: 273), italics original]

The biological species concept developed in the second half of the 19th century. Up to that time, from Plato and Aristotle until Linnaeus and early 19th century authors, one simply recognized "species," eide (Plato), or kinds (Mill). Since neither the taxonomists nor the philosophers made a strict distinction between inanimate things and biological species, the species definitions they gave were rather variable and not very specific. The word 'species' conveyed the idea of a class of objects, members of which shared certain defining properties. Its definition distinguished a species from all others. Such a class is constant, it does not change in time, all deviations from the definition of the class are merely "accidents," that is, imperfect manifestations of the essence (eidos). Mill in 1843 introduced the word 'kind' for species (and John Venn introduced 'natural kind' in 1866) and philosophers have since used the term natural kind occasionally for species... [(Mayr 1996: 266f) italics original]

Even though this ["the morphological, or typological species concept"] was virtually the universal concept of species, there were a number of prophetic spirits who, in their writings, foreshadowed a different species concept, later designated as the biological species concept (BSC). The first among these was perhaps Buffon (Sloan 1987), but a careful search through the natural history literature would probably yield quite a few similar statements. [*Op. cit.* 268]

The word interbreeding indicates a propensity; a spatially or chronologically isolated population, of course, is not interbreeding with other populations but may have the propensity to do so when the extrinsic isolation [is] terminated. [(Mayr 2000b: 17)]

Robert Sidney Bigelow (1918–2000)

Bigelow worked in New Zealand as a zoologist, specializing on grasshoppers (e.g., Bigelow 1967). His comment about gene flow has been widely cited.

... a bisexual species should be defined as: "a group of natural populations that is reproductively isolated from other such groups, but in which the component populations are not reproductively isolated from each other." Reproductive isolation [in Mayr's 1963 definition—*JSW*] should be considered in terms of gene flow, and not in terms of interbreeding, since selection will inhibit gene flow between well-integrated gene pools despite interbreeding. [(Bigelow 1965: 458)]

Hugh E. H. Paterson (1926–)

There have been numerous conceptions of species following Mayr that are fully, or partially, isolationist. The views of Hugh Paterson (1985; 1993) in particular have influenced Niles Eldredge (1989; 1993) and Elisabeth Vrba among others (Lambert and Spencer 1995). Paterson's version of the biospecies concept requires that organisms share a mating system, which he terms the "Specific-Mate Recognition System", or SMRS. He intends this to apply to plants, animals and other organisms, so the voluntarist implications of the term "recognition" should be taken as the term "selection" is, without voluntaristic or cognitive import. He refers to this as the Recognition Concept. Mayrian and Dobzhanskyan concepts he calls Isolation Concepts. Since Paterson's version applies, as did previous isolation concepts, only to fully sexual and gendered (anisogamous) organisms (p24), it follows that like them, he does not regard non-sexual organisms as forming species.

> *We can, therefore, regard a species as that most inclusive population of individual biparental organisms which share a common fertilization system.* [(Paterson 1985: 25, italics original)]

Alan R. Templeton (1946–)

Templeton considers two criteria for being a species—genetic and demographic exchangeability. The first he specifies as "the factors that define the limits of spread of new genetic variants through *gene flow*", and the second as "the factors that define the fundamental niche and the limits of spread of new genetic variants through *genetic drift* and *natural selection*" (see his table). He sees the cohesion concept as sharing a lot with the evolutionary species concept in this respect. These mechanisms generate a cohesive group, and the concept includes asexual taxa (purely in terms of demographic exchangeability) as well as sexual taxa (a mix of both). Nevertheless Templeton does not indicate by what the level of species is indicated. His is more a matter of identifying what mechanisms generate species, whatever that level may be. Under both mechanisms, the spread of genetic variants through populations is the *sine qua non* of species rank.

> The cohesion concept species is the most inclusive population of individuals having the potential for phenotypic cohesion through intrinsic cohesion mechanisms [(Templeton 1989: 12)]
>
> The cohesion concept of species defines a species as an evolutionary lineage through the mechanisms that limit the populational boundaries for the action of such basic microevolutionary forces as gene flow, natural selection, and genetic drift. The genetic essence of an evolutionary lineage is that a new mutation can go to fixation within it; and genetic drift and natural selection as well as gene flow are powerful forces that can cause such fixations. Hence, there is no good ratio-

nale for why gene flow should be the only microevolutionary mechanism that is used to define an evolutionary lineage; yet this is precisely what the isolation and recognition concepts do.

The "biological species concept" defines species as reproductive communities that are separated from other similar communities by intrinsic isolating barriers. However, there are other "biological" concepts of species, so the classic biological species concept is more accurately described as the "isolation" species concept. The purpose is this chapter was to provide a biological definition of species that follows directly from the evolutionary mechanisms responsible for speciation and their genetic consequences. [*Op. cit.* p20]

The strengths and weaknesses of the evolutionary, isolation, and recognition concepts were reviewed and all three were judged to be inadequate for this purpose. As an alternative, I proposed the cohesion concept that defines a species as *the most inclusive group of organisms having the potential for genetic and/or demographic exchangeability*. This concept borrows from all three biological species concepts. Unlike the isolation and recognition concepts, it is applicable to the entire continuum of reproductive systems observed in the organic world. Unlike the evolutionary concept, it identifies specific mechanisms that drive the evolutionary process of speciation. The cohesion concept both facilitates the study of speciation as an evolutionary process and is compatible with the genetic consequence of that process. [(Templeton 1989: 25), italics added]

James Mallet (1955–)

Mallet offers up "a species concept for the modern synthesis" (Mallet 1995; cf. also Mallet 2000; Mallet 2001; Dres and Mallet 2002; Naisbit et al. 2002; Beltran et al. 2002), the *genotypic cluster species concept* (GCSC) in which species were identified as "identifiable genotypic clusters" (Mallet 1995: 296). Although a genotypic notion, it in many ways resembles the phenetic concept in which instead of morphological variables, the lack of intermediates lies in single genetic locus and ensembles of multiple loci. Like the biospecies concept, the GCSC applies only to populations in sympatry or parapatry (Brower 2002), and lacks a nonarbitrary level of similarity or isolation of genetic alleles to specify specieshood.

An obvious alternative to the biological species concept is to define species in the darwinian way as distinguishable groups of individuals which have few or no intermediates when in contact, to extend the definition to cover polytypic species, and to incorporate new knowledge from genetics as well as morphology. When we observe a group of individuals within an area, we intuitively recognize species by means of morphology if there are no or few intermediates between two morphological clusters, and because independent characters that distinguish these clusters are correlated with each other. Adding genetics to this definition, we see two species rather than one if there are two identifiable genotypic clusters. These clusters are recognized by a deficit of intermediates, both at single loci (heterozygote deficits) and at multiple loci (strong correlations or disequilibria between loci which are divergent between clusters). Mendelian variation is discrete; therefore we expect quantized differences between individuals. We use the patterns of the discrete genetic differences, rather than the discreteness itself, to reveal genotypic clusters. [(Mallet 1995: 296)]

Species are recognized by morphological and genetic gaps between populations in a local area rather than by means of the phylogeny, cohesion, or reproductive isolation that are responsible for these gaps (Mallet 1995). *In a local area, a single species (the null hypothesis) is recognized if there is but a single cluster in the frequency distribution of multilocus phenotypes and genotypes. Separate species are recognized if there are several clusters separated by multilocus phenotypic or genotypic gaps.* [(Mallet 2001)]

🐍   Section 5. Evolutionary Conceptions

Evolutionary views of species treat a species taxon as a segment, as yet undivided, of the evolutionary tree. A basic concept required here is the concept of a *lineage*, which can mean a sequence of populations that give rise to each other, or a sequence of parent-child relations between organisms (see the notion of a "tokogenetic lineage" in Hennig's work), or even a sequence of genomes. "Lineage" is a substrate neutral concept, that can be applied whenever there are the requisite causal reproductive relations between predecessors and successors. De Queiroz has proposed a "Metapopulation Lineage Concept", in which a species is a sequence of lineages of a single metapopulation, for example. The notion of a lineage plays a crucial role in the work of philosopher David Hull (Hull 1981, 1984a) on the metaphysics of species.

Tracy Morton Sonneborn (1905–1981)

Sonnenborn took seriously the species concept in bacteria and other asexual organisms. As a consequence of taking evolution seriously, he proposed a notion of an evolutionary unit, the *syngen*. "Species" are a diagnostic entity.

> The nonexistence of species in asexual organisms is asserted only by those who define species as syngens. …
>
> Difficulties in the application of the term species arise from the attempt to make it do double duty in serving both as designating an evolutionary unit, the one which shows minimal irreversible discontinuity, and as designating a readily recognizable group. For most of the last hundred years, it has been tacitly assumed that these two aspects of species coincide and that, since the evolutionary unit represents a real level of biological organization, species exist as real entities, not as human constructs made for the convenience of biologists.
>
> … I propose, in the interest of logic and clarity, to use a different term, syngen, for the evolutionary unit. The same group of organisms would be both a species

and a syngen if both of the criteria of ready recognition and minimal irreversible evolutionary divergence were met. ...

Once the brave but hopeless effort of the proponents of the modern biological species concept is abandoned, it will again be generally recognized that species in sexual and asexual organisms have been and can be defined on essentially the same principle. That principle is simply minimal irreversible evolutionary divergence that yields readily recognizable difference. [(Sonneborn 1957: 289)]

George Gaylord Simpson (1902–1984)

Simpson, a paleontologist, is regarded as one of the founders of the Synthesis. In his comments regarding species, he introduces the notion of a "lineage" of populations, which became significant later, and he takes the evolutionary aspect of these lineages as crucial to understanding what species are (as Dobzhansky had earlier). Ghiselin (Ghiselin 1997: 112f) suggests the shift from the biospecies emphasis of his early definition to the less explicitly interbreeding concept a decade later is due to his falling "increasingly under the spell of the set-theoretical treatment of the Linnaean hierarchy by Gregg ..., who, although mentioning in passing the possibility that species are something else, insisted they are classes." According to Cain (1954: 111), Simpson characterized the intergrading forms of an evolutionary species as "transients". Species are thus a number of things: they are *populations* that form *phyletic lineages* through *interbreeding*, and which have *independent evolutionary roles* and *independent evolutionary tendencies*.

[A species is]... a phyletic lineage (ancestral-descendant sequence of interbreeding populations) evolving independently of others, with its own separate and unitary evolutionary role and tendencies, is a basic unit in evolution. [(Simpson 1951) (Quoted in Ghiselin 1997: 112f)]

An evolutionary species is a lineage (an ancestral-descendant sequence of populations) evolving separately from others and with its own unitary evolutionary role and tendencies. [(Simpson 1961: 153)]

Edward Orlando Wiley (1944–)

Wiley's version of the evospecies conception began life as a definition, but later becomes a "characterization". The novel elements in the later definition include the "entification" of evospecies in place of the population stipulation, to accommodate asexual organisms, and the inclusion of time and space to indicate the processual nature of the concept.

A species is a single lineage of ancestral descendant populations of organisms which maintains its identity from other such lineages and which has its own evolutionary tendencies and historical fate. [(Wiley 1978: 18)]

An evolutionary species is an entity composed of organisms that maintains its identity from other such entities through time and over space and that has its own independent evolutionary fate and historical tendencies. [(Wiley and Mayden 2000: 73)]

Evolutionary species are logical individuals with origins, existence, and ends. [*Op. cit.* p74]

Kevin de Queiroz (1956–)

De Querioz' view is titled the General Lineage Concept.

… a series of entities forming a single line of ancestry and descent. …

Lineages in the sense described above are unbranched; that is, they follow a single path or line anytime an entity in the series has more than one descendant. … Consequently, lineages are not to be confused with clades, clans, and clones— though the terms are often used interchangeably in the literature. (de Queiroz and Donoghue 1990: 50)]

✌ Section 6. Phylogenetic Conceptions

There are three kinds of phylospecies: Hennigian, Phylogenetic Taxon, and Autapomorphic. Hennigian species are lineal unbroken and unbranched segments of the evolutionary tree. Phylogenetic Taxon Conceptions treat species as the smallest group in a phylogenetic tree as exemplified by individual specimens, and Autapomorphic Conceptions treat species as diagnostic groups of unique characters.

As Nelson and Platnick note, Porphyry's Tree is strongly reminiscent of cladograms formed from the apomorphy/plesiomorphy distinction made by Willi Hennig (1966; 1950). Hennig distinguished between derived and underived states in taxonomy. *Plesiomorphic* characters are those in a monophyletic group (a clade) from which the transformations begin, and *apomorphies* are those derived from them in evolution (p. 89). An apomorphy of a group can be a plesiomorphy of a clade contained within that group. It follows that in both the neo-Platonic and Hennig's classificatory schemes that genus/plesiomorphy and species/apomorphy are not absolute ranks, but are relative. They are relative to the differentiae, the predicates and the properties those predicates denote which make the differences between taxa.

Hennig has no system of absolute taxonomic levels. In Gareth J. Nelson's view, species are just taxa, and they are arrayed in a flexible local hierarchy (Nelson 1989). The base-level taxonomic rank—the infimae species—was a taxic entity that was not itself the genus of any other species and which contained only individuals. Likewise, Hennig has terminal taxa, and these he calls species, following Linnaean tradition. Where the infimae species and the Hennigian species differ from the Linnaean species, however, is that the former are derived from the general group being sequentially divided into subgroups on the basis of characters shared (a single dichotomous key in the medieval conception, on a parsimony criterion of many characters for Hennig), while

Linnaeus assumed fixed taxon ranks. Linnaean species are an absolute rank, and so also are the higher taxa they comprise.

Species concepts based upon the phylogeny of the groups of organisms are called "phylogenetic", but there are several *phylospecies* concepts, as we will call them, and there are several sub-versions of them in turn. All proponents of phylospecies concepts claim both Darwin and Hennig as their inspirations, but it is arguable how closely the modern views relate to those initial expressions of classification of taxa by descent.

In some ways, phylogenetic classification is an outgrowth of the ideas expressed by Haeckel and others during the late nineteenth century. Haeckel coined the term *monophyly*, which Hennig later appropriated and more strictly defined. This is a critical notion in the context of phylogenetic systematics, or, as it is popularly known, cladistics. In Hennig's definition, a monophyletic group is a stem species and all of its descendent taxa. Under more formally defined notions of phylogeny, sometimes called pattern cladism, a monophyletic group is a proper subset of some set of taxa, without necessarily implying an immediate history. On this account, a monophyletic group is definable in terms of it having unique characters that are not shared by other taxa. These are called *apomorphies* in Hennig's terminology, or "derived characters". This is a relative term: an apomorphy for one set of taxa is a *plesiomorphy* (an underived, or primitive, character) for a more inclusive set of taxa. If one or more taxa share an apomorphy, it is called a *synapomorphy*, and if only a single taxon carries an apomorphy, it is called an *autapomorphy*.

There are fundamentally three phylospecies concepts. The first, defined initially by Hennig, is sometimes called the "Hennigian species concept" (Meier and Willmann 2000). It rests on a conventional decision Hennig made to be consistent with his broader philosophy of classification, and which we shall call the *Hennigian convention*. On this account, if a new species arises by splitting from a parental species, then we shall say that the parental species, no longer being monophyletic (that is, now being paraphyletic), has become extinct, and there are now two novel species.

The second phylospecies concept we might call the "Phylogenetic Taxon Concept" (PTC). It relies upon the notion that any monophyletic taxon is all the descendent lineages of a single stem lineage, and a species is an otherwise monophyletic taxon with no descendent lineages, but there are some ambiguities here. It is sometimes unclear whether a species is defined recursively as any taxon that has only a single ancestral species and no sister taxa that share that species, or whether a species is held to be any taxonomic lineage that has no taxonomic sub-lineages; in any event this phylogenetic unit concept is a hierarchical notion based upon history and biology.

The third phylospecies concept, which we shall call the "Autapomorphic Species Concept" (ASC),[1] a version of what we might term the Diagnostic Species Concept, is derived from the work of Donn Rosen (Rosen 1979). In various versions, it tends to rely upon the diagnosability of taxa (Cronquist 1978), or, as we may otherwise say, it is a largely epistemological notion of species. All the Autapomorphic concepts rely upon a species being the "terminal taxon" in a cladogram. Some proponents apply an evolutionary exegesis to this, while others restrict it to a diagnostic relationship, a distinction often referred to as the ontology-epistemology aspects of species (cf. Wheeler and Meier 2000). Of course, a phylospecies of one kind can also be a phylospecies of another. The PTC can play the same role as the ASC.

The division of the phylospecies into two main kinds, ignoring Hennigian species for the moment, is a reflection of the larger taxonomic debates. These raise the questions: (i) should classification proceed in terms of descent alone or on the basis of similarity (cladism versus gradism)? and (ii) if classification rests on clades, are homologies (apomorphies) indicators of history, or are patterns that are evidence in favor of a historical reconstruction but not themselves a model of evolution? Briefly, this is the distinction between ontological and epistemological notions of classification again.

Thos who take the epistemological classification position (grouping in terms of monophyly where that means homological relationships only) are generally those who fall under the now-moribund rubric of *pattern cladists*. Those who take the diagnosis of monophyly to give an immediate hypothesis of evolutionary history are the so-called orthodox, or "traditional" cladists, called "process cladists".[2] Process cladism tends to treat taxa as relationships between organisms, while pattern cladism tends to see taxa as composite entities of which organisms are members. In this regard, the traditional orthodoxy is

---

1   Meier and Willman (2000: 36-37) call the Autapomorphic Species Concept the Phylogenetic Species Concept *simpliciter*, and Davis (1997) calls the process or Phylogenetic Unit Concept the Autapomorphic Species Concept, in direct contradiction to the present usage. The terms shall be used as defined here, but it should be noted that there are other senses in the literature.

2   The term "process cladism" was introduced in print in Ereshefsky (2000) but was used previously in conversation. It is unclear to me that either pattern cladism or process cladism form monolithic schools, and the differences of opinion on these matters needs to investigated. Thanks to David Williams for noting this, and catching my transposition of their core ideas.

more closely allied to some aspects of the evolutionary species concepts of recent times.[3]

"Phylogenetic" approaches to species are founded on one of three criteria— synapomorphy, autapomorphy or prior phenetic identity as OTUs. Roughly, synapomorphy acts as the classical notion of genera, autapomorphy acts as the classical notion of differentia, and in the case of phenetic OTUs as inputs into a cladistic analysis, species here are types that are known prior to the formal division. Therefore, it is something of a misconstrual to think that there are only two "phylospecies" concepts besides the Hennigian account. In fact, there are several, and this binary division is a privative one; into diagnostic (Autapomorphic) species concepts, and all the other phylogenetic (cladistic) accounts.[4]

Operationally, phylogeneticists seem to have little difficulty in identifying the level of species taxa in their cladograms, but there is a problem that arises if the Autapomorphic Concept is taken too strictly. For instance, in a study of harbor seals (Burg, Trites, and Smith 1999), the authors have clearly already identified the species, and are considering whether or not the haplotype data taken from mitochondrial DNA support the claim that these populations form subspecies. But on a strict Autapomorphic Concept, *each* of the haplotype groups should be considered a separate species (in exactly the way Whately described a logical notion of *species* would for dog breeds), unless other considerations, such as the biogeography and interbreeding Burg *et al.* do use, are taken into account, and if they are, then the phylospecies concept is insufficient to delimit species. And so it appears that some sort of prior knowledge is required to specify at what level of a cladogram taxa begin and (for example) molecular lineages cease to be subspecific diagnostic criteria.

The claim that species are defined by constant characters, made by Wheeler and Platnick (2000), is problematic. Either we already know what characters count as species-defining, or we are unable to find a level of species (for there are constant characters for a great many higher level groups, as well as lower level groups, than the usual level at which species are identified). Either way, this

---

3   It is not, in my opinion, true that pattern cladism commits its adherents to an antievolutionary view of taxa. Neither is it true that it is an essentialistic view of taxonomy, as some, notably Mayr, have claimed. It is, however, typological in the way Farber (below) discusses.

4   Brent Mishler pointed out to me that what I had in earlier drafts been calling a single class of species concepts, under the term "Monophyletic Species Concepts" (the Phylogenetic Taxon Concept) was actually pretty diverse, and that some (e.g., Cracraft) did not think species had to be monophyletic.

is not a full conception of what makes a group a species. See in this respect, the claim made by Vrana and Wheeler (1992) below, under "Other Conceptions".

Emil Hans Willi Hennig (1913–1976)

In the work that began the cladistic revolution, and which remains the source for much of its thinking, so clearly and consistently was it expressed, Hennig treated species in an unusual manner (Hennig 1966: 28–32). He begins by defining "semaphoronts", or "character bearers" as the elements of systematics. These are effectively stages or moments in the lifecycle of organisms of the taxon (or rather, as characters are abstractions, they are representations of these stages). Individuals themselves, or more exactly the ontogenetic lifecycles of individuals, are related through reproductive relationships he called "tokogenetic" relationships. When tokogenetic relationships begin to diverge, they form species. Hennig is by implication asserting that species cease to exist when they are divided; that is, when the hologenetic relationships cease to be a single set (the species sets are represented in the diagram by the large ellipses). Although Hennig assumes that species are reproductive communities of harmoniously cooperating genes, as Dobzhansky had said, and that species are reproductive groups, as Mayr had said, he nevertheless assumes that species are phylogenetic lineages. In fact he says that species are defined over spatial dimensions as well, as "… a complex of spatially distributed reproductive communities, or if we call this relationship in space "vicariance," as a complex of vicarying communities of reproduction." [p47] Transformations of the species morphology and genetic composition within these two events do not affect the identity of the species, because the species is defined here as a homogenous reproductive community. In effect, Hennig is taking the biospecies isolationist concept to its limits. Species are extinguished at the next speciation event.

This has become known as the *Hennig Convention*. It has been strongly criticized from all sides, by biospecies proponents, evospecies proponents and other phylogeneticists (see the citations in Meier and Willmann 2000: 31), not least because it seems to be an arbitrary way to delimit species taxa. Mayr, for example, takes Hennig to be making a substantive claim on the ways species are formed at speciation, and criticizes it on the basis that the "parent" species can remain unchanged even though it is no longer monophyletic. But it seems to me that the critics have overlooked the most charitable interpretation of the Hennig Convention—it is a convention about *naming*. In short, the *name* of a species is extinguished at speciation. This follows from Hennig's views about the task of systematics. Using (and citing) Woodger and Gregg (see the excellent discussion in Wheeler and Platnick 2000) and the views of Woodger (1937) in particular about sets in classification, Hennig strives to ensure that there is no ambiguity of reference in the sets named in systematics. Since as soon as a set is divided there is ambiguity which of the two resultant sets is being referred to by a prior name, Hennig proposes to extinguish the now-ambiguous name and create two new ones. However, he seems to equivocate over whether or not they are new *entities* or not.

The Hennig account is fundamentally a biospecies concept. Hennig himself accepted that species were reproductively isolated, and the criteria used for identifying the relevant phylogenetic edges of the cladogram are simply those of the biospecies. We could therefore say that it is better considered to be considered under that rubric, and that the issue of "extinction" of species at cladogenesis is one of the reference of taxonomic names. In short, the "extinction" is a taxonomic extinction.

> When some of the tokogenetic relationships among the individuals of one species cease to exist, it disintegrates into two species and ceases to exist. It is the common stem species of the two daughter species. [(Hennig 1950: 102), Translated by Meier and Willman (2000: 30)]
>
> [Species arise]... when gaps develop in the fabric of the tokogenetic relationships. The genetic relationships that interconnect species we call phylogenetic relationships. The structural picture of the phylogenetic relationships differs as much from that of the individual tokogenetic relationships as the latter does from the structural picture of the ontogenetic relationships. In spite of these differences in their structural pictures, the phylogenetic, tokogenetic, and ontogenetic relationships are only portions of a continuous fabric of relationships that interconnect all semaphoronts and groups of semaphoronts. With Zimmermann we will call the totality of these the "hologenetic relationships." [(Hennig 1966): 30]
>
> The limits of the species in the longitudinal section through time would consequently be determined by two processes of speciation: the one through which it arose as an independent reproductive community, and the other through which the descendants of this initial population ceased to exist as a homogenous reproductive community. [*Op. cit.* p58]

Rudolph Meier (1918–), and Rainer Willman (1950–)

The Hennigian concept of species has been recently expanded and defended by several people. Meier and Willmann (Meier and Willmann 2000: 31; cf. Willmann 1985b; 1985a; 1997) have proposed a modified Hennigian Species

Concept. Like Hennig, and the proponents of the biospecies and other isolationist conceptions, they reject the use of a single taxonomic category such as *species* to apply to asexual ("uniparental") organisms. Instead, they call them "agamotaxa".

> Species are reproductively isolated natural populations or groups of natural populations. They originate via the dissolution of the stem species in a speciation event and cease to exist either through extinction or speciation. [(Quoted from Willmann 1985a: 80, 176)]

Joel Cracraft (1943–), and Niles Eldredge (1943–)

Cracraft and Eldredge proposed the first of the Phylogenetic Taxon Conceptions in which species were not evolutionary units, but phylogenetic entities that did not map well onto evolutionary species. Cracraft subsequently revised his formulation by removing mention of reproductive cohesion ("like kind")

> We conclude … that the strict application of cladistic methodology to the problem of species recognition is not entirely consistent with the concept of species as individuals and evolutionary units. Use of patterns of synapomorphy to delineate monophyletic groups requires that all taxa be terminal, whereas the very notion of evolution requires some unit—and it must be the species— serve as ancestors and descendants. Strict adherence to cladistic methodology may tend to underestimate the true number of species sampled. A corollary, and somewhat ironic, observation, is that species are not always monophyletic units in a cladistic sense. [Niles Eldredge and Joel Cracraft, italics original (1980: 90)]

> … *a species is a diagnosable cluster of individuals within which there is a parental pattern of ancestry and descent, beyond which there is not, and which exhibits a pattern of phylogenetic ancestry and descent among units of like kind.* [*Ibid*, italics original (Eldredge and Cracraft 1980: 92)]

> A species is the smallest diagnosable cluster of individual organisms within which there is a parental pattern of ancestry and descent. [Joel Cracraft (1983: 170)]

Brent Drennen Mishler (1953–), Robert N. Brandon (1952–), Michael J. Donoghue, and Edward Claiborne Theriot (1953–)

Mishler and Brandon (1987) summarize their Monophyletic version (Mishler and Donoghue 1982) of phylospecies. Contrary to Cracraft and Eldredge, they do require monophyly of species, ruling out multiple speciation events causing a single species. They redefine *monophyly* in such a way as to be able to include species.

De Queiroz and Donoghue, on the other hand, treat species as systems that may not be monophyletic, and indeed may be paraphyletic if a species has split from it, in a parallel with cohesive and functional individuals who lose cells and reproduce (de Queiroz and Donoghue 1988, 1990). They therefore exclude asexuals and indistinct populations that are not assignable from being members of species. Mishler and Theriot (2000), extending the monophyletic conception, include asexuals (p52f) largely on the grounds that asexual taxa do not markedly differ in overall phylogenetic nature from sexuals—the number of autapomorphies, for example, are similar in both cases (one does wonder, though, if this is due more to the application of phylogenetic classification than anything else).

Their PTC is based on the historical, actual, lines of ancestry and descent that are represented in a cladogram. As a phylogenetic taxon, a species is grouped by the synaporphies shared by organisms that indicates monophyly (Mishler and Theriot 2000: 47). It is a phylogenetic or cladistic replacement for the evolutionary species concept of Simpson.

De Queiroz and Donoghue (1988; 1990), however, do not think that species have to be monophyletic, because monophyly of populations does not offer a way to specify what the base rank is, and because species evolve from ancestral populations this will leave the species from which the ancestral population derived as paraphyletic. Of course, the monophyly spoken of here is somewhat different from the monophyly of the Mishler, *et al.* Version; this one is based on populations as the base entities; theirs is based on phylogenetic lineages. And both conceptions converge on a similar solution—species are regarded as singular phylogenetic *lineages*, which is de Queiroz's later conception of a cohesive object or group over phylogeny (de Queiroz 1998; 1999, see above). The actual answer of the earlier paper, however, is that there is no single definition of species that will "answer to the needs of all biologists and will be applicable to all organisms" (quoting Kitcher 1984: 309), although they reject the sort of limited pluralism proposed, for example, by Wilkins (2003).

A species is the least inclusive taxon recognized in a classification, into which organisms are grouped because of evidence of monophyly (usually, but not restricted to, the presence of synapomorphies), that is ranked as a species because it is the smallest 'important' lineage deemed worthy of formal recognition, where 'important' refers to the action of those processes that are dominant in producing and maintaining lineages in a particular case. [(Mishler and Brandon 1987;: 310 in Hull and Ruse 1998)]

A monophyletic taxon is a group that contains all and only descendants of a common ancestor, originating in a single event. [*Op. cit.* (p313 in Hull and Ruse 1998)]

A species is the least inclusive taxon recognized in a formal phylogenetic classification. As with all hierarchical levels of taxa in such a classification, organisms are grouped into species because of evidence of monophyly. Taxa are ranked as species because they are the smallest monophyletic groups deemed worthy of formal recognition, because of the amount of support for their monophyly and/or because of their importance in biological processes operating on the lineage in question. [Mishler and Theriot (2000: 46–47)]

Arthur John Cronquist (1919–1992)

Diagnosis of species has always played a key role in the debate, but few if any have until recently suggested that diagnosis is sufficient, apart from (and possibly not even there) the Taxonomic Species Concept. Cronquist (1978: 3) provided one of the first such conceptions. Some (e.g., Ghiselin 1997: 106f) accuse Cronquist of presenting a "subjective" concept, but it all hinges on what "ordinary means" means. At one time, the use of a microscope was reviled (e.g., by Linnaeus); now assays for specificity ranging from molecular data to morphometric and acoustic traits are considered ordinary practice.

Species are the smallest groups that are consistently and persistently distinct, and distinguishable by ordinary means. [Cronquist (1978: 3)]

Donn Eric Rosen (1929–1986)

More common diagnostic concepts, though, arise from the recognition that what makes taxa distinct are their apomorphies. Species have diagnostic autapomorphies—that is to say, they have unique constellations of characters— while higher taxa (clades) have *syn*apomorphies—shared constellations of characters, which group them together. One instance of this approach is found in the work of Donn Rosen, whose definitions are the original source of the Autapomorphic Conception, and influence Nelson and Platnick's subsequent definition and thinking. He then noted that this means that subspecies are, "by definition, unobservable and undefineable", since they have no such apomorphies.

[A species is] … a geographically constrained group of individuals with some unique apomorphous characters, is the unit of evolutionary significance. [(Rosen 1978: 176; quoted in Wheeler and Platnick 2000: 55)]

… a species is merely a population or group of populations defined by one or more apomorphous features; it is also the smallest natural aggregation of individuals with a specifiable geographic integrity that cannot be defined by any set of analytic techniques. [(Rosen 1979: 277; quoted in Mayr 2000a: 99)]

Gareth J. Nelson (1937–), Norman I. Platnick (1951–), and Ward Christopher Wheeler (1963–)

The 1981 Nelson and Platnick definition, I am informed by Professor Nelson, was not intended to be a formal definition, but instead a passing comment describing current practice. It is one of the Autapomorphic Conceptions. They, too, noted that this meant that diagnosable "subspecies" were thus species.

The Wheeler and Platnick variation is prior to a cladistic analysis, and so like Nelson's earlier suggestion that species are taxa like any other level of a phylogenetic tree (Nelson 1989), they do not worry about characters, apomorphies and homologies when recognizing species. Species are found wherever characters are fixed and constant across all samples, while traits may be variable. Their willingness to handle and include asexual taxa within their species concept marks them out from most other phylospecies conceptions, and they bite the bullet on recognizing clones of asexual lineages as species. Mishler considers a number of the diagnostic accounts to be phenetically based, including Cracraft's, Platnick's and Nixon and Wheeler's (pers. comm.). Whether this is so (that is, whether they make use of the Cartesian clustering of species in a state space of traits typical of phenetic practice), it is clear that they assume that species are phylogenetically speaking the terminal taxa on a tree. Diagnosis assumes that the traits are specified before the tree is constructed.

[A species is] … simply the smallest detected samples of self-perpetuating organisms that have unique sets of characters. [(Nelson and Platnick 1981: 12; quoted in Wheeler and Platnick 2000: 56)]

We define species as the smallest aggregation of (sexual) populations or (asexual) lineages diagnosable by a unique combination of character traits. This concept represents a unit concept. [(Wheeler 1999; Wheeler and Platnick 2000): 58]

If the goal of distinguishing species is thereby to recognize the end-products of evolution, should we seek to suppress naming large numbers of species where large numbers of differentiated end-products exist? [*Idem*]

Phylocode

The Phylocode, a proposal for replacing the Linanean system of classification, has no definition of species beyond a generic diagnostic conception.

species

A segment of a population-level lineage that is evolving separately from other such lineage segments as indicated by one or more lines of evidence (e.g., distinguishability, reproductive isolation, monophyly, etc.). [(Cantino and de Querioz 2000), glossary]

ઢ Section 7. Ecological Conceptions

Michael Tenant Ghiselin (1939–)

The most extensive units in the natural economy such that reproductive competition occurs among its parts. [(Ghiselin 1974b: 38)]

Leigh Maiorana Van Valen (1935–)

A species is a lineage (or a closely related set of lineages) which occupies an adaptive zone minimally different from that of any other lineage in its range and which evolves separately from all lineages outside its range.

A lineage is a clone or an ancestral–descendent sequence of populations. A population is a group of individuals in which adjacent individuals at least occasionally exchange genes with each other reproductively, and in which adjacent individuals do so more frequently than with individuals outside the population.

Lineages are closely related if they have occupied the same adaptive zone since their latest common ancestor. If their adaptive zone has changed since then, they are closely related if the new adaptations have been transferred among the lineages rather than originating separately in each. [(Van Valen 1976: 233)]

[An adaptive zone is] Some part of the resource space together with whatever predation and parasitism occurs on the group considered. [*Op. cit.* p234]

[A "multispecies" is] … set of broadly sympatric species that exchange genes in nature… [*Op. cit.* p235]

❧ Section 8. Asexual Conceptions

Arthur James Cain (1921–1999)

Fresh from a stint working with Mayr, Cain wrote the enormously influential undergraduate text *Animal Species and their Evolution*, which set out the terms of the subsequent debate in detail. In the course of this, he accepts Turesson's a term for asexual species, *agamospecies*, and identifies them as "Morphological Species".

> It seems advisable to recognize clearly the limits of the biological species, and it is convenient to distinguish those forms to which it cannot apply because they have no true sexual reproduction as *agamospecies*. Species-criteria for such forms are the same as ... for the morphological species. [(Cain 1954: 103)]

Manfred Eigen (1927–)

A Nobel laureate for physics, Eigen turned his attention to the origin of life and the nature of asexual groups. He devised a notion for viruses that he called "quasispecies", which were a cluster of genomes in a genomic hyperspace, maintained by selection for the optimal genome for infecting the host cells. The "wild-type" is either the mode of this cluster, or the ancestor of all subsequent clonal mutants.

> A viral species ... is actually a complex, self-perpetuating population of diverse, related entities that act as a whole [(Eigen 1993: 32)]
>
> [A "quasispecies" is] ... region in sequence space [which] can be visualized as a cloud with a center of gravity at the sequence from which all mutations arose. It is a self-sustaining population of sequences that reproduce themselves imperfectly but well enough to retain a collective identity over time. [*Op. cit.* p35]

Verne Edwin Grant (1917–2007)

Grant, as a botanist, observed the need for asexual species early on, given the number of self-pollinating plant species.

> The reversion to asexual reproduction, as numerous authors have pointed out, spells the end of species in the biological sense of the word. The smallest integrated unit above the individual is the clone or biotype; the smallest well-defined taxonomic unit may be a huge polymorphic complex, an agamic or clonal complex.
>
> For purposes of classification, the taxonomist must name and describe selected morphological types in asexual groups. Whether these aggregations of individuals should be called "species" or not is a matter of taste. One tradition favors a general and hence indiscriminate use of the category of species in all groups of organisms. Some botanists would qualify this usage in asexual groups by employing the terms *agamospecies* or *taxonomic species*. The term *agameon* has also been suggested [(Camp and Gilly 1943)].
>
> The present author prefers the neutral term, *binom*, suggested by Camp [(1951)]. The grounds for this preference are that it saves the term species for the isolated population system and avoids the confusion of concepts inherent in the application of the same word to two basically dissimilar phenomena. [(Grant 1957: 61)]

⚘ Section 9. Other Conceptions

Species deniers or species replacers deny that the usual connotation of "species" has any biological reality, and that the real unit of biology is the organism. Taxa are abstractions of properties of groups of organisms, that are diagnosed, not discovered. Some species deniers offer terms designed to replace "species" that are rank-neutral.

Peter Henry Andrews Sneath (1923–), and Robert Reuven Sokal (1926–)

Sneath and Sokal developed the phenetic approach to classification, and instead of species argued in favour of similarity-based Operational Taxonomic Units, (OTUs) defined as clusters of traits with a 70% similarity or greater.

> In the absence of data on breeding and in apomictic groups …., the species are based on the phenetic similarity between the individuals and on phenotypic gaps. These are assumed to be good indices of the genetic position, although they need not be. … In this book it [the term "species"] will be used in the sense of phenetic rank … . [(Sokal and Sneath 1963: 35)]

Frederick Pleijel, and Greg W. Rouse

Pleijel, whose work focuses on the taxonomy of velvet worms, has denied that the term "species" is necessary, and that the operative taxonomic unit is the least taxonomic unit.

> ["Least Inclusive Taxonomic Units" are] … named monophyletic groups which are identified by unique shared similarities (apomorphies) … which are at present not further subdivided. … Identification of taxa as LITUs are statements about the current state of knowledge (or lack thereof) without implying that they have no internal nested structure; we simply do not know if a given LITU consists of several monophyletic groups or not. [Pleijel and Rouse (2000: 629)]

Paul B. Vrana, and Ward Christopher Wheeler (1963–)

Vrana and Wheeler reject the necessity for a particular taxonomic rank as species, arguing that taxonomists only need to make reference to individual organisms.

It is therefore obvious (by the emphasis on sex within definitions) that there is a bias on the part of species definitions by authors towards multi-cellular sexually reproducing groups. We believe this is undesireable, and that any such theory should extend to all organisms. The notion that asexual organisms constitute an insignificant minority unworthy of consideration in a general theory of biology is based on lack of information, given that few systemic studies focus on protozoan, bacterial or viral groups. Wheeler ... as well as de Queiroz and Donoghue ... have agreed that any number of unicellular asexual organisms which show pattern may be termed the equivalent of species. If one uses this not-unreasonable definition (given that this taxon is not seen as a group of organisms or tokogeny = phylogeny) and the realization that each of the "sexual species" hosts at least several of the "asexual"—protozoan, fungal, bacterial and viral—then it must be the former that seem insignificant and aberrant in general biological mechanisms. Some may attempt to portray sex as a more general mechanism, extending to many of the "asexual species". It should not be glibly assumed that the occasional (often one-way) exchange of genetic material in these organisms is equivalent to sexual reproduction in metazoans. This is tantamount to stating that retroviral infection of a human host is that person having sex with the virus. Espouse such a definition if you must, but be aware of the ramifications.

Certainly such processes should play no part in the delimitation of the smallest taxa possible. In practice, every systematist examines individual organisms. Yet, the unit used in the systematic analysis is the species. When there is variation among individuals this is either ignored (common = primitive) or coded as ambiguous—an artificial suite of characters not necessarily observed in any real organism. These seem to us to be the best reasons for using individuals as terminal units in phylogenetic analysis. If there is polymorphism within species in characters that one would like to compare among species, perhaps one should consider establishing the monophyly of the group first. If a number of individuals are identical for all characters considered, they could then be considered one taxon for the purposes of phylogenetic analysis (this may lessen practical objections to our view, as it renders the most parsimonious cladogram more obtainable). We realize that such a species definition, that is as arbitrary as any other level, has disturbing implications for biodiversity. Casting doubt on the validity of the units being counted renders the conclusions of these studies somewhat dubious.

Our view does, however, make cladistic analysis relevant to population genetics, a field riddled with untestable assumptions (and responsible for many of the molecular evolution paradigms), as well as recent biogeographic events ... and conservation biology. As an example of the latter, one may apply cladistic analysis to a number of organisms belonging to a taxon known to be endangered ("species", "genus" or whatever) and obtain a cladogram showing structure between less inclusive taxa of these organisms. A defensible strategy of conservation can then be devised, based on the principle of maximizing cladistic diversity ... . Thus, in our view cladistics is a more generally applicable method of analysis.

The cladistically determined hierarchy of individuals implies a picture of genealogy—the categorical rank of species is irrelevant. Which, if any, of the groups revealed by genealogy is a species is an entirely different question. We suggest no change in terminology, however; the term species may still be applied to groups on the basis of interfertility. Indeed, this may be appropriate for those explicitly pursuing questions of process; we simply ask that such groups not be used in phylogenetic analysis. As Nelson ... has said: "there seem to be no basic taxonomic unit and no particular unit of evolution" and (species) "problems are insoluble, for they stem from a false assumption: that there is an empirical difference between species and other taxa". In keeping with this, we feel that use of individual organisms as terminal entities is the only logical choice. [(Vrana and Wheeler 1992: 70f)]

The Essentialism Story developed gradually at first, in the comments about the role played by Aristotle and scholasticism in natural history, in particular in Dewey's essay on Darwin and philosophy below. But where we might expect essence of species to play a role in the criticism of prior thought, particularly among the Darwinians in the period from 1860 to 1900, we don't. Instead we find criticisms of the nature of investigation, or epistemological assumptions, such as in the *History of Botany* by enthusiastic Darwinian Julius Sachs, the first edition of which was published shortly into the Darwinian period in 1875. If we were to find this contrast, we might expect it here of all places. Instead, we find criticism of the Aristotelian program of doing science by definition. W. Stanley Jevons, whose 1873 discussion on classification acts as a nice segue from the older period to the modern, notes that classes have to have definienda common to all members of the class, but that classification in biology actually isn't classification at all, but an arrangement of groups, not classes, by genealogy. This is the earliest claim in the philosophical literature that classification in biology is not really about classes, and appears to be a reflection upon the fact that taxonomy was about types, which are individuals not classes, in the light of evolution.

As is often noted, the Darwinian revolution was as much an epistemological or methodological shift as it was a theoretical one (Ghiselin 1984, 2005; Ruse 1999). After the development of set theory, however, a distinction of the scholastics between intension and extension, of sets that were circumscribed by definitions or by member inclusion, was revived, and the logical tradition of species was held to be a matter of definition. In a seminal summary of the traditional pre-set theoretic logic of *diairesis*, H. W. Joseph (1916) made a clear distinction, as Whately had in 1826, between logical species and "natural" species, but the developers of the Essentialism Story failed to pick this up, and read him as saying that species of living things were the same as the logical species (Chung 2003; Winsor 2001, 2003). Around the time of the Centenary of the publication of the *Origin*, in 1959, this story began to be developed, and it was elaborated over the subsequent 20 years, until it became the received view in the history and philosophy of biology.

In part motivated by this story, and in part motivated by developments in philosophy, David Hull published a paper which was titled "The Effects of Essentialism on Taxonomy: Two Thousand Years of Stasis" (Hull 1965), in which the claim was implicitly made that fixity of species was due to Aristotelian influences. However, fixism was motivated by piety and theology, not philosophy, and essentialism in logic was, if anything, a post hoc justification of that claim. Or it would be if there were strong historical evidence that essentialism in natural species was a widespread view. And typology, as Winsor argued, was ubiquitous as a "method of exemplars" (Winsor 2003), and it remains the standard approach even today, by Darwin and the evolutionists as much as by the idealists or fixists.

Dissent from the Essentialism story began fairly early. The passage below from Paul Farber indicates that "type" had a more complex and nuanced role in taxonomy in the period immediately before Darwin than the Essentialism Story recounted. And there is no reason to think that any of these senses had any essentialist implications. Stung by criticisms of being "typological", some cladists began to investigate the historical material, and argued similarly (Nelson and Platnick 1981), and others (Atran 1985, 1990; Stevens 1994) followed.

Hull's paper was followed by a seminal paper by Michael Ghiselin, in which he argued that species were not classes or natural kinds, but logical individuals (Ghiselin 1974a), a view shortly after endorsed and expanded by Hull (Hull 1976). I stress that the arguments presented in these papers do not rely for their *philosophical* force on the Essentialism Story being historically correct. They are responding to Popper and other views (such as Strawson, and later, Kripke) that were current.

Finally, I include some modern quotations that I trust are sufficiently refuted by Darwin's writings in section 3, that Darwin did not know what species were, or denied they existed.

William Stanley Jevons (1835–1882)

Jevons is remarkable in many respects, as the thinker who first proposed both a logical calculus with quantifiers ("some" or "all") but who also built a "logic piano" which would, like a calculator, generate logic inferences. However his magisterial *Principles of Science*, first published in 1873, indicates some of the more deliberative thinking about classification in the light of evolution. Jevons more than anyone else was in a position to marry the older logic, for the work of Boole was still largely unknown and undigested, with the new biology.

His is the last real philosophical discussion of biological classification for many years. Several of his claims are significant in this context: that classification in biology was of a different kind to classification in mineralogy, chemistry and physics, a "special" kind; that genealogy rather than affinity was the basis for the best botanical and zoological classifications, that groups in biology were formed by the use of types rather than definitions (see above, Whewell and Mill) because there were likely to be exceptions to any definitional differentiae; and that for other purposes (largely to do with utility) classification by correlations of properties was better. He notes that a "class must be defined by the invariable presence of certain properties" and that classification by types is in fact a shorthand of a complex system of arranging organisms. The type, he says, "itself is an individual, not a class", which suggests that for him natural species are individuals.

Jevons was not, however, a species realist. Like most of his contemporaries, he thought that the theory of evolution necessitated species conventionalism, or "nominalism", in which the degree of variation was arbitrarily divided into groups such as species by naturalists. There is no criterion of specific and generic difference, and so there need to be "an unlimited series of subaltern classes". Hence, he notes, the newly adopted Linnaean "laws" of classification by de Candolle, involve up to twenty-one ranks, with the threat of more. Intriguingly, in the light of Atran's work (Atran 1985, 1990, 1995, 1998, 1999), he notes that Linnaeus thought too there was something psychologically compelling about five ranks. He concludes with the comment "that natural classification in the animal and vegetable kingdoms is a special problem [of classification in science], and that the particular methods and difficulties to which it gives rise are not those common to all cases of classification, as so many physicists have supposed. Genealogical resemblances are only a special case of resemblances in general."

Classification in the Biological Sciences.

The great generalisations established in the works of Herbert Spencer and Charles Darwin have thrown much light upon other sciences, and have removed several difficulties out of the way of the logician. The subject of classification has long been studied in almost exclusive reference to the arrangement of animals and plants. Systematic botany and zoology have been commonly known as the Classificatory Sciences, and scientific men seemed to suppose that the methods of arrangement, which were suitable for living creatures, must be the best for all other classes of objects. Several mineralogists, especially Mohs, have attempted to arrange minerals in genera and species, just as if they had been animals capable of reproducing their kind with variations. This confusion of ideas between the relationship of living forms and the logical relationship of things in general prevailed from the earliest times, as manifested in the etymology of words. We familiarly speak of a kind of things meaning a class of things, and the kind consists of those things which are akin, or come of the same race. When Socrates and his followers wanted a name for a class regarded in a philosophical light, they adopted the analogy in question, and called it a γένος, or race, the root γέν- being connected with the notion of generation.

So long as species of plants and animals were believed to proceed from distinct acts of Creation, there was no apparent reason why methods of classification suitable to them should not be treated as a guide to the classification of other objects generally. But when once we regard these resemblances as hereditary in their origin, we see that the sciences of systematic botany and zoology have a special character of their own. There is no reason to [719] suppose that the same kind of natural classification which is best in biology will apply also in mineralogy, in chemistry, or in astronomy. The logical principles which underlie all classification are of course the same in natural history as in the sciences of lifeless matter, but the special resemblances which arise from the relation of parent and offspring will not be found to prevail between different kinds of crystals or mineral bodies.

The genealogical view of the relations of animals and plants leads us to discard all notions of a regular progression of living forms, or any theory as to their symmetrical relations. It was at one time a question whether the ultimate scheme of natural classification would lead to arrangement in a simple line, or a circle, or a combination of circles. Macleay's once celebrated system was a circular one, and each class-circle was composed of five order-circles, each of which was composed again of five tribe-circles, and so on, the subdivision being at each step into five minor circles. Macleay held that in the animal kingdom there are five sub-kingdoms—the Vertebrata, Annulosa, Radiata, Acrita, and Mollusca. Each of these was again divided into five—the Vertebrata, consisting of Mammalia, Reptilia, Pisces, Amphibia, and Aves.[1] It is evident that in such a symmetrical system the animals were made to suit themselves to the classes instead of the classes being suited to the animals.

We now perceive that the ultimate system will have the form of an immensely extended genealogical tree, which will be capable of representation by lines on a plane surface of sufficient extent. Strictly speaking, this genealogical tree ought to represent the descent of each individual living form now existing or which has existed. It should be as personal and minute in its detail of relations, as the Stemma of the Kings of England. We must not assume that any two forms are exactly alike, and in any case they are numerically distinct. Every parent then must be represented at the apex of a series of divergent lines, representing the generation of so many children. Any complete system of classification must regard individuals as the infimae species. But as in the lower races of animals

[1] Swainson, "Treatise on the Geography and Classification of Animals," *Cabinet Cyclopaedia*, p. 201.

[720] and plants the differences between individuals are slight and apparently unimportant, while the numbers of such individuals are immensely great, beyond all possibility of separate treatment, scientific men have always stopped at some convenient but arbitrary point, and have assumed that forms so closely resembling each other as to present no constant difference were all of one kind. They have, in short, fixed their attention entirely upon the main features of family difference. In the genealogical tree which they have been unconsciously aiming to construct, diverging lines meant races diverging in character, and the purpose of all efforts at so-called natural classification was to trace out the descents between existing groups of plants or animals.

Now it is evident that hereditary descent may have in different cases produced very different results as regards the problem of classification. In some cases the differentiation of characters may have been very frequent, and specimens of all the characters produced may have been transmitted to the present time. A living form will then have, as it were, an almost infinite number of cousins of various degrees, and there will be an immense number of forms finely graduated in their resemblances. Exact and distinct classification will then be almost impossible, and the wisest course will be not to attempt arbitrarily to distinguish forms closely related in nature, but to allow that there exist transitional forms of every degree, to mark out if possible the extreme limits of the family relationship, and perhaps to select the most generalised form, or that which presents the greatest number of close resemblances to others of the family, as the *type* of the whole.

Mr. Darwin, in his most interesting work upon Orchids, points out that the tribe of Malaxeae are distinguished from Epidendreae by the absence of a caudicle to the pollinia; but as some of the Malaxeae have a minute caudicle, the division really breaks down in the most essential point. "This is a misfortune," he remarks,[1] "which every naturalist encounters in attempting to classify a largely developed or so-called natural group, in which, relatively

¹ Darwin, *Fertilisation of Orchids*, p. 159.

[721] to other groups, there has been little extinction. In order that the naturalist may be enabled to give precise and clear definitions of his divisions, whole ranks of intermediate or gradational forms must have been utterly swept away: if here and there a member of the intermediate ranks has escaped annihilation, it puts an effectual bar to any absolutely distinct definition."

In other cases a particular plant or animal may perhaps have transmitted its form from generation to generation almost unchanged, or, what comes to the same result, those forms which diverged in character from the parent stock may have proved unsuitable to their circumstances, and perished. We shall then find a particular form standing apart from all others, and marked by many distinct characters. Occasionally we may meet with specimens of a race which was formerly far more common but is now undergoing extinction, and is nearly the last of its kind. Thus we explain the occurrence of exceptional forms such as are found in the Amphioxus. The Equisetaceae perplex botanists by their want of affinity to other orders of Acrogenous plants. This doubtless indicates that their genealogical connection with other plants must be sought for in the most distant ages of geological development.

Constancy of character, as Mr. Darwin has said,¹ is what is chiefly valued and sought after by naturalists; that is to say, naturalists wish to find some distinct family mark, or group of characters, by which they may clearly recognise the relationship of descent between a large group of living forms. It is accordingly a great relief to the mind of the naturalist when he comes upon a definitely marked group, such as the Diatomaceae, which are clearly separated from their nearest neighbours the Desmidiaceae by their siliceous framework and the absence of chlorophyll. But we must no longer think that because we fail in detecting constancy of character the fault is in our classificatory sciences. Where gradation of character really exists, we must devote ourselves to defining and registering the degrees and limits of that gradation. The ultimate natural arrangement will often be devoid of strong lines of demarcation.

¹ *Descent of Man*, vol. i. p. 214.

[722] Let naturalists, too, form their systems of natural classification with all care they can, yet it will certainly happen from time to time that new and exceptional forms of animals or vegetables will be discovered and will require the modification of the system. A natural system is directed, as we have seen, to the discovery of empirical laws of correlation, but these laws being purely empirical will frequently be falsified by more extensive investigation. From time to time the notions of naturalists have been greatly widened, especially in the case of Austra-

lian animals and plants, by the discovery of unexpected combinations of organs, and such events must often happen in the future. If indeed the time shall come when all the forms of plants are discovered and accurately described, the science of Systematic Botany will then be placed in a new and more favourable position, as remarked by Alphonse Decandolle.[1]

It ought to be remembered that though the genealogical classification of plants or animals is doubtless the most instructive of all, it is not necessarily the best for all purposes. There may be correlations of properties important for medicinal, or other practical purposes, which do not correspond to the correlations of descent. We must regard the bamboo as a tree rather than a grass, although it is botanically a grass. For legal purposes we may continue with advantage to treat the whale, seal, and other cetacea, as fish. We must also class plants according as they belong to arctic, alpine, temperate, sub-tropical or tropical regions. There are causes of likeness apart from hereditary relationship, and *we must not attribute exclusive excellence to any one method of classification.*

Classification by Types.
Perplexed by the difficulties arising in natural history from the discovery of intermediate forms, naturalists have resorted to what they call classification by types. Instead of forming one distinct class defined by the invariable possession of certain assigned properties, and rigidly including or excluding objects according as they do or do not

[1] *Laws of Botanical Nomenclature*, p. 16.

[723] possess all these properties, naturalists select a typical specimen, and they group around it all other specimens which resemble this type more than any other selected type. "The type of each genus," we are told,[1] "should be that, species in which the characters of its group are best exhibited and most evenly balanced." It would usually consist of those descendants of a form which had undergone little alteration, while other descendants had suffered slight differentiation in various directions.

It would be a great mistake to suppose that this classification by types is a logically distinct method. It is either not a real method of classification at all, or it is merely an abbreviated mode of representing a complicated system of arrangement. A class must be defined by the invariable presence of certain common properties. If, then, we include an individual in which one of these properties does not appear, we either fall into logical contradiction, or else we form a new class with a new definition. Even a single exception constitutes a new class by itself, and by calling it an exception we merely imply that this new class closely resembles that from which it diverges in one or two points only. Thus in the definition of the natural order of Rosaceae, we find that the seeds are one or two in each carpel, but that in the genus Spiraea there are three or four; this must mean either that the number of seeds is not a part of the fixed definition of the class, or else that Spiraea does not belong to that class, though it may closely approximate to it. Naturalists continually find themselves between two horns of a dilemma; if they restrict the number of marks specified in a definition so that every form intended to come within the class shall possess all those marks, it will then be usually found to include too many forms; if the definition be made more particular, the result is to produce so-called anomalous genera, which, while they are held to belong to the class, do not in all respects conform to its definition. The practice has hence arisen of allowing considerable latitude in the definition of natural orders. The family of Cruciferae, for instance, forms an exceedingly well-marked natural order, and among its characters we find it

[1] Waterhouse, quoted by Woodward in his *Rudimentary Treatise of Recent and Fossil Shells*, p. 61.

[724] specified that the fruit is a pod, divided into two cells by a thin partition, from which the valves generally separate at maturity; but we are also informed that, in a few genera, the pod is one-celled, or indehiscent, or separates transversely into several joints.[1] Now this must either mean that the formation of the pod is not an essential point in the definition of the family, or that there are several closely associated families.

The same holds true of typical classification. The type itself is an individual, not a class, and no other object can be exactly like the type. But as soon as we abstract the individual peculiarities of the type and thus specify a finite number of qualities in which other objects may resemble the type, we immediately constitute a class. If some objects resemble the type in some points, and others in other points, then each definite collection of points of resemblance constitutes intensively a separate class. The very notion of classification by types is in fact erroneous in a logical point of view. The naturalist is constantly occupied in endeavouring to mark out definite groups of living forms, where the forms themselves do not in many cases admit of such rigorous lines of demarcation. A certain laxity of logical method is thus apt to creep in, the only remedy for which will be the frank recognition of the fact, that, according to the theory of hereditary descent, gradation of characters is probably the rule, and precise demarcation between groups the exception.

Natural Genera and Species.

One important result of the establishment of the theory of evolution is to explode all notions about natural groups constituting separate creations. Naturalists long held that every plant belongs to some species, marked out by invariable characters, which do not change by difference of soil, climate, cross-breeding, or other circumstances. They were unable to deny the existence of such things as sub-species, varieties, and hybrids, so that a species of plants was often subdivided and classified within itself. But then the differences upon which this sub-classification [725] depended were supposed to be variable, and thus distinguished from the invariable characters imposed upon the whole species at its creation. Similarly a natural genus was a group of species, and was marked out from other genera by eternal differences of still greater importance.

We now, however, perceive that the existence of any such groups as genera and species is an arbitrary creation of the naturalist's mind. All resemblances of plants are natural so far as they express hereditary affinities; but this applies as well to the variations within the species as to the species itself, or to the larger groups. All is a matter of degree. The deeper differences between plants have been produced by the differentiating action of circumstances during millions of years, so that it would naturally require millions of years to undo this result, and prove experimentally that the forms can be approximated again. Sub-species may sometimes have arisen within historical times, and varieties approaching to sub-species may often be produced by the horticulturist in a few years. Such varieties can easily be brought back to their original forms, or, if placed in the original circumstances, will themselves revert to those forms; but according to Darwin's views all forms are capable of unlimited change, and it might possibly be, unlimited reversion if suitable circumstances and sufficient time be granted.

Many fruitless attempts have been made to establish a rigorous criterion of specific and generic difference, so that these classes might have a definite value and rank in all branches of biology. Linnaeus adopted the view that the species was to be defined as a distinct creation, saying,[1] "Species tot numeramus, quot diversae formae in principio sunt createa;" or again, "Species tot sunt, quot diversas formas ab initio produxit Infinitum Ens; quae formae, secundum generationis inditas leges, produxere plures, at sibi semper similes." Of genera he also says,[2] "Genus omne est naturale, in primordio tale creatum." It was a common doctrine added to and essential to that of distinct creation that these species could not produce intermediate and variable forms, so that we find Linnaeus obliged by the ascertained existence of hybrids to take a different view

[1] *Philosophia Botanica* (1770), §157, p. 99.

[2] *Ibid.* §159, n. 100.

[726] in another work ; he says,[1] "Novas species immo et genera ex copula diversarum specierum in regno vegetabilium oriri primo intuitu paradoxum

videtur; interim observationes sic fieri non ita dissuadent." Even supposing in the present day that we could assent to the notion of a certain number of distinct creational acts, this notion would not help us in the theory of classification. Naturalists have never pointed out any method of deciding what are the results of distinct creations, and what are not. As Darwin says,[2] "the definition must not include an element which cannot possibly be ascertained, such as an act of creation." It is, in fact, by investigation of forms and classification that we should ascertain what were distinct creations and what were not; this information would be a result and not a means of classification.

Agassiz seemed to consider that he had discovered an important principle, to the effect that general plan or structure is the true ground for the discrimination of the great classes of animals, which may be called branches of the animal kingdom.[3] He also thought that genera are definite and natural groups. "Genera," he says,[4] "are most closely allied groups of animals, differing neither in form, nor in complication of structure, but simply in the ultimate structural peculiarities of some of their parts; and this is, I believe, the best definition which can be given of genera." But it is surely apparent that there are endless degrees both of structural peculiarity and of complication of structure. It is impossible to define the amount of structural peculiarity which constitutes the genus as distinguished from the species.

The form which any classification of plants or animals tends to take is that of an unlimited series of subaltern classes. Originally botanists confined themselves for the most part to a small number of such classes. Linnaeus adopted Class, Order, Genus, Species, and Variety, and even seemed to think that there was something essentially natural in a five-fold arrangement of groups.[5]

[1] *Amoenitates Academicae* (1744), vol. i. p. 70. Quoted in *Edinburgh Review*, October 1868, vol. cxxviii. pp. 416, 417.

[2] *Descent of Man*, vol. i. p. 228.

[3] Agassiz, *Essay on Classification*, p. 219.

[4] *Ibid.* p. 249.

[5] *Philosophia Botanica*, §155, p. 98.

[727] With the progress of botany intermediate and additional groups have gradually been introduced. According to the Laws of Botanical Nomenclature adopted by the International Botanical Congress, held at Paris[1] in August 1867, no less than twenty-one names of classes are recognised—namely, Kingdom, Division, Sub-division, Class, Sub-class, Cohort, Sub-cohort, Order, Sub-order, Tribe, Sub-tribe, Genus, Sub-genus, Section, Sub-section, Species, Sub-species, Variety, Sub-variety, Variation, Sub-variation. It is allowed by the authors of this

scheme, that the rank or degree of importance to be attributed to any of these divisions may vary in a certain degree according to individual opinion. The only point on which botanists are not allowed discretion is as to the order of the successive sub-divisions; any inversion of the arrangement, such as division of a genus into tribes, or of a tribe into orders, is quite inadmissible. There is no reason to suppose that even the above list is complete and inextensible. The Botanical Congress itself recognised the distinction between variations according as they are Seedlings, Half-breeds, or *Lusus Naturae*. The complication of the inferior classes is increased again by the existence of *hybrids*, arising from the fertilisation of one species by another deemed a distinct species, nor can we place any limit to the minuteness of discrimination of degrees of breeding short of an actual pedigree of individuals.

It will be evident to the reader that in the remarks upon classification as applied to the Natural Sciences, given in this and the preceding sections, I have not in the least attempted to treat the subject in a manner adequate to its extent and importance. A volume would be insufficient for tracing out the principles of scientific method specially applicable to these branches of science. What more I may be able to say upon the subject will be better said, if ever, when I am able to take up the closely-connected subjects of Scientific Nomenclature, Terminology, and Descriptive Representation. In the meantime, I have wished to show, in a negative point of view, that natural classification in the animal and vegetable kingdoms is a special problem, and that the particular methods and

¹ *Laws of Botanical Nomenclature*, by Alphonse Decandolle, translated from the French, 1868, p. 19.

[728] difficulties to which it gives rise are not those common to all cases of classification, as so many physicists have supposed. Genealogical resemblances are only a special case of resemblances in general.

[From *Principles of Science*, second edition, 1878, bk V. chapter XXX (Jevons 1878: 718–728)]

John Dewey (1859–1952)

Dewey is a member of the pragmatist school of American philosophy which was greatly influenced by Darwin's theory, moreso than most philosophical schools. In this essay, he sets up a strong contrast between post-Darwinian thought and pre-Darwinian thought that is, I think, unjustified.

Few words in our language foreshorten intellectual history as much as does the word species. The Greeks, in initiating the intellectual life of Europe, were impressed by characteristic traits of the life of plants and animals; so impressed indeed that they made these traits the key to defining nature and to explaining mind and society. And truly, life is so wonderful that a seemingly successful reading of its mystery might well lead men to believe that the key to the secrets of heaven and earth was in their hands. The Greek rendering of this mystery, the Greek formulation of the aim and standard of knowledge, was in the course of time embodied in the word species, and it controlled philosophy for two thousand years. True investigation of nature consists not only in deducing rules from exact and comparative observation of the phenomena of nature, but in discovering the genetic forces from which the causal connexion, cause and effect may be derived. In pursuit of these objects, it is compelled to be constantly correcting existing conceptions and theories, producing new conceptions and new theories, and thus adjusting our own ideas more and more to the nature of things. The understanding does not prescribe to the objects, but the objects to the understanding. The Aristotelian philosophy and its medieval form, scholasticism, proceeds in exactly the contrary way; it is not properly concerned with acquiring new conceptions and new theories by means of investigation, for conceptions and theories have been once and for all established; experience must conform itself to the ready-made system of thought; whatever does not so conform must be dialectically twisted and explained till it apparently fits in with the whole. (Sachs 1890: 86)

To understand the intellectual face-about expressed in the phrase "Origin of Species," we must, then, understand the long dominant idea against which it is a protest.

Consider how men were impressed by the facts of life. Their eyes fell upon certain things slight in bulk, and frail in structure. To every appearance, these perceived things were inert and passive. Suddenly, under certain circumstances, these things—henceforth known as seeds or eggs or germs—begin to change, to change rapidly in size, form, and qualities. Rapid and extensive changes occur, however, in many things as when wood is touched by fire. But the changes in the living thing are orderly; they are cumulative; they tend constantly in one direction; they do not, like other changes, destroy or consume, or pass fruitless into wandering flux; they realize and fulfil. Each successive stage, no matter how unlike its predecessor, preserves its net effect and also prepares the way for a fuller activity on the part of its successor. In living beings, changes do not happen as they seem to happen elsewhere, any which way; the earlier changes are regulated in view of later results. This progressive organization does not cease till there is achieved a true final term, τελὸς, a completed, perfected end. This final form exercises in turn a plenitude of functions, not the least note worthy of which is production of germs like those from which it took its own origin, germs capable of the same cycle of self-fulfilling activity.

But the whole miraculous tale is not yet told. The same drama is enacted to the same destiny in countless myriads of individuals so sundered in time, so severed in space, that they have no opportunity for mutual consultation and no means of interaction. As an old writer quaintly said, "things of the same kind go through the same formalities"—celebrate, as it were, the same ceremonial rites.

This formal activity which operates throughout a series of changes and holds them to a single course; which subordinates their aimless flux to its own perfect manifestation; which, leaping the boundaries of space and time, keeps individuals distant in space and remote in time to a uniform type of structure and function: this principle seemed to give insight into the very nature of reality itself. To it Aristotle gave the name, εἶδος. This term the scholastics translated as *species*.

The force of this term was deepened by its application to everything in the universe that observes order in flux and manifests constancy through change. From the casual drift of daily weather, through the uneven recurrence of seasons and unequal return of seed time and harvest, up to the majestic sweep of the heavens—the image of eternity in time—and from this to the unchanging pure and contemplative intelligence beyond nature lies one unbroken fulfilment of ends. Nature as a whole is a progressive realization of purpose strictly comparable to the realization of purpose in any single plant or animal.

The conception of εἶδος, species, a fixed form and final cause, was the central principle of knowledge as well as of nature. Upon it rested the logic of science. Change as change is mere flux and lapse; it insults intelligence. Genuinely to know is to grasp a permanent end that realizes itself through changes, holding them thereby with in the metes and bounds of fixed truth. Completely to know is to relate all special forms to their one single end and good: pure contemplative intelligence. Since, however, the scene of nature which directly confronts us is in change, nature as directly and practically experienced does not satisfy the conditions of knowledge. Human experience is in flux, and hence the instrumentalities of sense-perception and of inference based upon observation are condemned in advance. Science is compelled to aim at realities lying behind and beyond the processes of nature, and to carry on its search for these realities by means of rational forms transcending ordinary modes of perception and inference.

There are, indeed, but two alternative courses. We *must* either find the appropriate objects and organs of knowledge in the mutual interactions of changing things; or else, to escape the infection of change, we must seek them in some transcendent and supernal region. The human mind, deliberately as it were, exhausted the logic of the changeless, the final, and the transcendent, before it essayed adventure on the pathless wastes of generation and transformation. We dispose all too easily of the efforts of the schoolmen to interpret nature and mind in terms of real essences, hidden forms, and occult faculties, forgetful of the seriousness and dignity of the ideas that lay behind. We dispose of them by laughing at the famous gentleman who accounted for the fact that opium put people to sleep on the ground it had a dormitive faculty. But the doctrine, held in our own day, that knowledge of the plant that yields the poppy consists in referring the peculiarities of an individual to a type, to a universal form, a doctrine so firmly established that any other method of knowing was conceived to be unphilosophical and unscientific, is a survival of precisely the same logic. This identity of conception in the scholastic and anti-Darwinian theory may well suggest greater sympathy for what has become unfamiliar as well as greater humility regarding the further unfamiliarities that history has in store.

Darwin was not, of course, the first to question the classic philosophy of nature and of knowledge. The beginnings of the revolution are in the physical science of the sixteenth and seventeenth centuries. When Galileo said: "It is my opinion that the earth is very noble and admirable by reason of so many and so different alterations and generations which are incessantly made therein," he expressed the changed temper that was coming over the world; the transfer of interest from the permanent to the changing. When Descartes said: "The nature of physical things is much more easily conceived when they are beheld coming gradually into existence, than when they are only considered as produced at once in a finished and perfect state," the modern world became self-conscious of the logic that was henceforth to control it, the logic of which Darwin's "Origin of Species" is the latest scientific achievement. Without the methods of Copernicus, Kepler, Galileo, and their successors in astronomy, physics, and chemistry, Darwin would have been helpless in the organic sciences. But prior to Darwin the impact of the new scientific method upon life, mind, and politics, had been arrested, because between these ideal or moral interests and the inorganic world intervened the kingdom of plants and animals. The gates of the garden of life were barred to the new ideas; and only through this garden was there access to mind and politics. The influence of Darwin upon philosophy resides in his having conquered the phenomena of life for the principle of transition and thereby freed the new logic for application to mind and morals and life. When he said of species what Galileo had said of the earth, *e pur si muove*, he emancipated, once for all, genetic and experimental ideas as an organon of asking questions and looking for explanations. [John Dewey, "The Influence of Darwin on Philoso-phy", 1910 (Dewey 1997: 4–7)]

John Henry Woodger (1894–1981)

Woodger, and John Gregg (Gregg 1954), attempted to formalise classification in terms of the new formal logic of Tarski and Frege. In so doing, Woodger sets up a distinction between abstract classes and concrete particulars that contributes to the later debates on the metaphysics of species.

> In the Linnean system of classification of animals and plants a species was a set or class, in fact it originally meant a smallest named class in the system. But a class or set is an abstract entity and thus has neither beginning nor end in time. We cannot, therefore, speak of the origin of species if we are conceiving species in the Linnean manner. The doctrine of evolution is not something that can be grafted, so to speak, onto the Linnean system of classification. The species of Darwin and the species of Linneus are not at all the same thing—the former are concrete entities with a beginning in time and the latter are abstract and timeless. [(Woodger 1952: 19)]

Ernst Walter Mayr (1904–2005)

Mayr introduced the notion that the Greek *eidos* was an abstraction too, at first. But this is a Platonic view, read through Lockean eyes. He also claimed that Darwin was a species denier.

> Typological thinking finds it easy to reconcile the observed variability of the individuals of a species with the dogma of the constancy of species because the variability does not affect the essence of the eidos [the Greek term translated as "species"] which is absolute and constant. Since the eidos is an abstraction derived from human sense impressions, and a product of the human mind, according to this school, its members feel justified in regarding a species "a figment of the imagination," an idea. [(Mayr 1957: 12)]
>
> … one might get the impression [from the *Origin of Species*] that he considered species as something purely arbitrary and invented merely for the convenience of taxonomists. [(Mayr 1982: 268)]

Morton O. Beckner (1928–2001)

Beckner's work is perhaps the first of the modern philosophies of biology. In it, he discussed whether species had a set of necessary and defining attributes or

were instead a disjunct of properties, possession of most of which would suffice
to include an organism in the species.

> [Taxa are not well defined (W-defined) but are, or at least could be, effectively or
> empirically defined (E-defined)] ... if a species is E-defined at all, it is polytypic
> with respect to the properties utilized in the description. If the species-concept
> is not fully polytypic, the description provides genuine diagnostic characters-
> in-common; but it should be noticed that the systematist never, or at any rate
> seldom, knows that his species-concept is not fully polytypic, and I suspect many
> of the systematist's polymorphic species and Rassenkreise are. [(Beckner 1959:
> 65)]

George Gaylord Simpson (1902–1984)

Simpson's comment here is symptomatic of the increasing interest in
philosophical contributions to the species concept in biology, after the Centenary
of the *Origin*. Here he is clearly referring to the "ideal morphologists" of the
early 19th century and later, who were themselves Platonists.

> The basic concept of typology is this: every natural group of organisms, hence
> every natural taxon in classification, has an invariant, generalized or idealized
> pattern shared by all members of the group. The pattern of the lower taxon
> is superimposed upon that of the higher taxon to which it belongs, without
> essentially modifying the higher pattern. ... Numerous different terms have
> been given to these idealized patterns, often simply "type" but also "archetype,"
> "Bauplan" or "structural plan," "Morphotypus" or "morphotype," "plan" and
> others. [(Simpson 1961: 54)]

Michael Tenant Ghiselin (1939–)

Ghislein is the originator of the Individuality Thesis—that species are logical
individuals. Later he moves further to claiming that species are also functional
individuals, which is to say that the populations form a cohesive gene pool that
tracks ecological niches.

> A universal is simply a general term, or concept, such as "the state". Nominalists
> would deny the existence of the state, but affirm the existence of individual
> states. What of species? Nominalists would deny the existence of the species, but
> affirm the existence of individual species ... Biological species are, in the logical
> sense, individuals, and assertions to the contrary reflect mere verbal confusion.
> [(Ghiselin 1966: 208f)]

There are both advantages and disadvantages to treating species as individuals. … such a position has attractive qualities from the point of view of logic and biology alike, and … it is perhaps not so radical as one might think. The basic point … is that multiplicity does not suffice to render an entity a mere class. In logic, "individual" is not a synonym for "organism". Rather, it means a particular thing. It can designate systems at various levels of integration. … An individual … need not be physically continuous. … It is characteristic of individuals that there cannot be instances of them. … individuals do not have "intensions." That is to say, there are no properties necessary and sufficient to define their names. [(Ghiselin 1974a: 536f)]

Paul Lawrence Farber (1944–)

Farber is the first critic of the identification of "type" with "essence" by the Essentialist Story, although he is not directly attacking that story so much as clarifying the use of the term "type" by naturalists in the early 19th century. This paper is worth quoting at length for its redefinition of the problem.

The type-concept currently is in ill repute. For the most part, it is regarded as a theoretical component of pre-Darwinian natural science that presented a major obstacle to the development and acceptance of modern evolutionary ideas. This perspective, from the vantage of contemporary biology, overlooks the important role of the type-concept and oversimplifies its meaning in the first half of the nineteenth century. During that period, it functioned as a central organizing idea in zoology. Moreover, it was not a simple notion. Rather, it was a constellation of concepts that zoologists employed in different specialties, assigned to different levels of organization, and interpreted in different ways. [(Farber 1976: 93)]

In a simple form, the type-concept had long been used by naturalists as an aid in classification and nomenclature. We may call this type-concept the *classification type-concept*. [*Ibid* p93]

A second meaning of the type-concept emerged from the professional zoology done in museum and other large collections during the first half of the nineteenth century. The research of the workers in these collections dealt with either classification or morphology. Their efforts to organize the ever-swelling empirical base of zoological data led them to attempt to put some order into their own method. Especially pressing was the problem of distinguishing new species, and from this problem emerged the recognition of the need for an accurate identification of the particular specimen used by an author to name a new species. Type-specimens came to be carefully labeled and became a valuable part of major collections because they could be consulted as reference material for

re-examination at a later date. Each served as a model and name carrier by embodying the distinctive characteristics of the species. An individual belonged to
the same species if, when compared with the type-specimen, it did not display
any essential differences. The type-specimen was not a perfect guide since it was
of value mostly for external morphology, which was not always the sole criterion
for identification, however, it was of greater value than a plate or written description of these same characters.

Similar to the classification type-concept, then, the *collection type-concept*
served as a model and name carrier. It differed from the former, however, in that
it referred to a particular specimen in a known collection rather than to a general group. [*Ibid* p96f]

Morphology, perhaps more than any other discipline, focused attention on the
type-concept, for comparative anatomists were interested in a basic plan or type
that they believed could be discerned at various taxonomic levels. Like the classification type-concept, the morphological type-concept was used implicitly before the nineteenth century. [*Ibid* p100]

More important than the state of the type-specimen, however, is that naturalists, who wished to treat type-specimens as typical specimens, were confusing
the collection type-concept with the morphological or classification type-concept. It is obvious that an average specimen is more desirable as a type-specimen than one from the far range of variation; nonetheless, one specimen could
never be more typical than another in the sense of embodying the morphological
type, which was a plan of organization that in principle consisted of essential elements. Nor could the type-specimen be typical in the sense of serving as a model
in the classification type-concept. A type-species served as the model for a genus
and allowed naturalists to define other species by noting their essential similarities *and* differences, whereas a type-specimen was an individual that served as
a model to establish that another individual either had or did not have essential
similarities. [*Ibid* p107]

The type-concept had its negative side as well as its positive side. A major
weakness of the morphological type-concept was its ambiguity concerning lower
taxa, especially species. The species level was critical because the great increase
of specimens from all over the world that became available to naturalists for
study resulted in the number of recognized species increasing dramatically. With
this increase came considerable confusion about how to distinguish true species from sub-species, rarities, races, and so on. A naturalist might be able to
describe a specimen accurately, but how was he to decide if a variation in color
was important enough to justify naming a new species? What about geographi-

cal variation? Behavioral differences? The "lumper-splitter" debate is, of course, a perennial problem, but the failure of comparative anatomists precisely to define the morphological type-concept was a particular handicap in dealing with the mass of material available because the concept implied that one could define exactly the characteristics of species when in fact one was unable practically to supply precise criteria. [*Ibid* p118]

David Lee Hull (1935–)

Hull was the first philosopher to adopt Ghiselin's Individuality Thesis, and he elaborated its philosophical implications at length. Here the notion that to be a class is to be intensionally defined by essential properties is attacked. Later he despairs of there being a single species concept that will satisfy everyone.

> On the traditional view, the species category is a class of classes defined in terms of the properties which particular species possess ... and particular organisms are individuals ... The relation between organisms, species and the species category is membership. An organism is a member of its species and each species is a member of the species category. On the view being urged [by Ghiselin and Hull], both particular species and the species category must be moved down one category level. Organisms remain individuals, but they are no longer members of their species. Instead an organism is part of a more inclusive individual, its species, and the names of both particular organisms (like Gargantua) and particular species (like *Gorilla gorilla*) become proper names. The species concept is no longer a class of classes but merely a class. [Hull (1976: 174f)]

> One reason why philosophers find the monism-pluralism debate so interesting is its apparent connection to the dispute over realism and antirealism. Of the four possible combinations of these philosophical positions, two seem quite natural: monism combined with realism, and pluralism combined with antirealism. ...

> The other two combinations ... are somewhat strained. It would seem a bit strange to argue that one and only one way exists to divide up the world, but that groups of natural phenomena produced on this conceptualisation are not "real". They are as real as anything can get! ... A combination of pluralism and realism seems equally peculiar ... . The world can be divided up into kinds in numerous different ways, and the results are all equally real! [(Hull 1999: 25)]

Douglas Futuyma (1942–)

Futuyma's textbook has served generations of evolutionary biology undergraduates.

> One of the ironies of the history of biology is that Darwin did not really explain the origin of new species in *The Origin of Species*, because he didn't know how to define a species. [(Futuyma 1983: 152)]

Michael Ruse (1940–)

Ruse, a close friend and sparring partner of Hull's, is one of the few current philosophers to argue for a class-view of species.

> A perhaps more fruitful way of tackling the problem of the ontological status of species is to examine precisely what is entailed in calling something an "individual" rather than a class. Here is surely the reason why speaking of a species as being like a super-organism does not strike most of us as so stupid, even if we decide in the end that that is not quite what we want to claim. For the point is that individuals have some kind of integration, some kind of internal ordering principle, which non-individuals do not have. … The question which faces us now therefore is whether a species has the kind of internal integration, cohesive structure, functioning, that we associate with individuals of a more traditional kind. … The science has passed by the species-as-individuals thesis. … it just did not work. … Back to classes, I think … [Ruse (1998: 286f)]

Jody Hey (1958–)

Hey, a Rutgers geneticist, argues that species do not exist and that they are just mind-relative categories.

> The presumption that species do exist, and that we will try to avoid, actually has two parts. The first and most obvious is that organisms occur as part of distinct real entities that we might call species, and that every organism belongs to precisely one such real entity. The second common presumption is the idea that species have something in common. [(Hey 2001: 68)]

John Simpson Wilkins (1955–)

Wilkins is claiming that the concept of species has something in common in all its applications, and that the "essential" definition is that of lineages of descent held separate from other such lineages. Nevertheless, he accepts that there are different definitions or conceptions of "species" that depend on the particular group of organisms under study, and their actual mechanisms of separation.

> *A species is a lineage separated from other lineages by causal differences reflected in synapomorphies.*
>
> ...
>
> *Species are lineages of organisms that have differences in shared traits that keep them distinct from other lineages.*
>
> ...
>
> *Species are lineages of populations with homologies that are differently expressed, and which prevent the populations from recombining.*
>
> [(Wilkins 2003), italics original]

Jerry A. Coyne (1949–), H. Allen Orr (1960–)

In a major review of research into speciation, Jerry Coyne and Allen Orr present a revised version of the BSC in which they make a number of concessions to criticisms and in which they defend against some (Coyne and Orr 2004, chapter 1). Coyne's and Orr's version, which we will here call the *limited BSC*, allows for limited introgression, does not insist upon integration of gene complexes, permits other definitions for asexuals and paraphyly of species, and treats ecological differentiation as necessary to the persistence of species (in sympatry, at any rate) but not to the definition based on reproductive isolation. It allows that non-genetic isolation can form species (for example, intracellular infection by *Wollbachia*), and further that good species might later hybridize to form a new species. They accept Mayr's definition: "species are groups of interbreeding natural populations that are reproductively isolated from other such groups" (Mayr 1995: 5), although given their qualifying of that concept, it should perhaps read **mostly** interbreeding and isolated.

> ... *The Origin of Species*, whose title and first paragraph imply that Darwin will have much to say about speciation. Yet his magnum opus remains largely silent on the "mystery of mysteries," and the little it does say about this mystery is seen by most modern evolutionists as muddled or wrong. [(Coyne and Orr 2004: 9)]
>
> ... biologists want species concepts to be useful for some purpose (i.e., be "operational"), but differ in what that purpose should be. We can think of at least five such goals. Species can be defined in a way that
>
> 1. helps us classify them in a systematic manner;
> 2. corresponds to the discrete entities that we see in nature;
> 3. helps us understand how discrete entities *arise* in nature;
> 4. represents the evolutionary history of organisms; and

5. applies to the largest possible number of organisms.

No species concept will accomplish even most of these purposes. We therefore feel that, when deciding on a species concept, one should first identify the nature of one's "species problem," and then choose the concept best at solving that problem.

...

Our own species concept is one that comes closest to deciphering what we (and many of our predecessors) consider the most important "species problem," namely, why do sympatric, sexually reproducing organisms fall into discrete clusters? This view of the species problem antedates the Modern Synthesis, going back to Bateson (1894). In our opinion, the discontinuities of nature are best encapsulated, and their origin best understood, using a modified version of the biological species concept... [p26]

In our view, distinct species are characterized by *substantial but not necessarily complete reproductive isolation*. We thus depart from the "hard line" BSC by recognizing species that have limited gene exchange with sympatric relatives. But we feel that it is less important to worry about species status than to recognize that the *process* of speciation involves acquiring reproductive barriers, and that this process yields intermediate stages when species status is more or less irresolvable. [p30]

Scott Atran (1952–)

Atran is most concerned with the anthropological and conceptual aspects of living kind classification, but he does take some time to argue against the Essentialism Story. As such he is perhaps one of the earliest explicit attackers of the story.

For my part, I have so far failed to find any natural historian of significance who ever adhered to the strict version of essentialism so often attributed to Aristotle. Nor is any weaker version of the doctrine that has indiscriminately been attributed to Cesalpino, Ray, Tournefort, A.-L. Jussieu and Cuvier likely to bear up under closer analysis. [(Atran 1990: 85)]

Peter Francis Stevens (1944–)

Stevens' excellent work on Jussieu also dealt with the development of species as isolated taxonomic classes. He notes that the Great Chain view of Linnaeus ended up with Jussieu a map of isolated "territories" and agrees with Farber's view of types.

Collection type concepts were concerned with how the name of an organism can be referred to a particular specimen or individual species. Classification type concepts were those that dealt largely with summarizing or simplifying data, whereas morphological concepts dealt with the order of nature and its laws (although Paul Farber, whose work is followed here, noted that these two were not always sharply distinguishable) [Stevens (1994: 134)]

Joel Cracraft (1943–)

Cracraft, an ornithologist at the AMNH, and author of a phylogenetic conception of species, also has been one of the more sensible commentators on the species debate. I will give him the last word.

> The literature on species concepts has always contained a high proportion of philosophical ruminations by biologists and has received considerable attention by philosophers. The biologists generally are not very sophisticated as philosophers and the philosophers are not very sophisticated as biologists—but that is all right. At least there has been a dialogue, and it has sharpened and informed the discussion and controversy in useful ways.

> Take the notions of reality, individuality, and entities. All these figure prominently in recent discussions about species concepts... But one has to exercise caution because as with all discussions about reality, it pays to be aware of my sixth law of thermodynamics, never falsified: things aren't as they seem. (Cracraft 2000: 10–11)

> ... somehow a species definition must be inclusive of an ontology and an epistemology. [Joel Cracraft (2000: 13)]

❧   Section 11. A List of 26 Species Definitions in the Modern Literature[1]

There are numerous species "concepts" at the research and practical level in the scientific literature. (Mayden 1997) list of 22 distinct species conceptions along with synonyms is a useful starting point for a review. I have added authors where I can locate them in addition to Mayden's references, and I have tried to give the concepts names, such as *biospecies* for "biological species", and so on (following George 1956), except where nothing natural suggests itself. There have also been several additional conceptions since Mayden's review, which I have added the views of Pleijel (1999) and Wu (2001b; Wu 2001a), and several new revisions presented in Wheeler and Meier (2000). I also add some "partial" species conceptions—the *compilospecies* conception and the *nothospecies* conception.

Asterisks identify the "basic" or "elemental" conceptions, from which the others are formed.

Phylospecies

In addition to Hennig's conception [nr 15], I distinguish between two *phylospecies* concepts that go by various names, mostly the names of the authors presenting at the time (as in Wheeler and Meier 2000). To remedy this terminological inflation, I have christened them the *Autapomorphic species conception* [nr 2] and the *Phylogenetic Taxon species conception* [nr 21].

A phylospecies is the smallest unit appropriate for phylogenetic analysis, the smallest biological entities that are diagnosable, the unit product of natural selection and descent, a geographically constrained group with one or more unique apomorphies (autapomorphies). There are two versions of this and they are not identical. One derives from Rosen and is what I call the

---

1   Reprinted with amendments from (Wilkins 2006).

Autapomorphic species conception. It is primarily a concept of diagnosis and tends to be favored by the tradition known as pattern cladism. The other is what I call the *Phylogenetic Taxon species conception*, and tends to be favored by process cladists. See Cracraft (1983); Eldredge and Cracraft (1980); Nelson and Platnick(1981); Rosen (1979). *Partial synonyms*: Autapomorphic phylospecies, monophyletic phylospecies, minimal monophyletic units, monophyletic species, lineages.

1. Agamospecies*

Asexual lineages, uniparental organisms (parthenogens and apomicts), that cluster together in terms of their genome. May be secondarily uniparental from biparental ancestors. Quasispecies are asexual viruses or organisms that cluster about a "wild-type" due to selection. See Cain (1954), Eigen (1993, for quasispecies). *Synonyms*: Microspecies, binoms, paraspecies, pseudospecies, semispecies, quasispecies, genomospecies (for prokaryotes Euzéby: 2006).

2. Autapomorphic species

A phylospecies conception. A geographically constrained group of individuals with some unique apomorphous characters, the unit of evolutionary significance, Rosen (1979); simply the smallest detected samples of self-perpetuating organisms that have unique sets of characters, Nelson and Platnick (1981); the smallest aggregation of (sexual) populations or (asexual) lineages diagnosable by a unique combination of character traits (Wheeler and Platnick 2000).

3. Biospecies*

Defined by John Ray, Buffon, Dobzhansky (1935); Mayr (1942). Inclusive Mendelian population of sexually reproducing organisms (Dobzhansky 1935, 1937, 1970), interbreeding natural population isolated from other such groups (Mayr 1942, 1963, 1970; Mayr and Ashlock 1991). Depends upon endogenous reproductive isolating mechanisms (RIMs). *Synonyms*: Syngen, speciationist species concept.

4. Cladospecies

Set of organisms between speciation events or between speciation event and extinction (Ridley 1989), a segment of a phylogenetic lineage between nodes.

Upon speciation the ancestral species is extinguished and two new species are named. See Hennig (1966; 1950); Kornet (1993). *Synonyms*: Internodal species concept, Hennigian species concept, Hennigian convention

5. Cohesion species

Evolutionary lineages bounded by cohesion mechanisms that cause reproductive communities that are demographically (ecologically) interchangeable. See Templeton (1989).

6. Compilospecies

A species pair where one species "plunders" the genetic resources of another via introgressive interbreeding. See Harlan (1963), Aguilar (1999).

7. Composite Species

All organisms belonging to an internodon and its descendents until any subsequent internodon. An internodon is defined as a set of organisms whose parent-child relations are not split (have the INT relation). See Kornet (1993).

8. Ecospecies*

A lineage (or closely related set of lineages) which occupies an adaptive zone minimally different from that of any other lineage in its range and which evolves separately from all lineages outside its range. See Simpson (1961); Sterelny (1999); Turesson (1922); Van Valen (1976). *Synonyms*: Ecotypes

9. Evolutionary species*

A lineage (an ancestral-descendent sequence of populations) evolving separately from others and with its own unitary evolutionary role and tendencies. See Simpson (1961); Wiley (1978); (1981). *Synonyms*: Unit of evolution, evolutionary group, syngen (partial). *Related concepts*: Evolutionary significant unit.

10. Evolutionary significant unit

A population (or group of populations) that (1) is substantially reproductively isolated from other conspecific population units, and (2) represents an important component in the evolutionary legacy of the species. See Waples (1991).

11. Genealogical concordance species

Population subdivisions concordantly identified by multiple independent genetic traits constitute the population units worthy of recognition as phylogenetic taxa. See Avise and Ball (1990).

12. Genic species

A species formed by the fixation of all isolating genetic traits in the common genome of the entire population. See Wu (2001b; 2001a).

13. Genetic species*

Group of organisms that may inherit characters from each other, common gene pool, reproductive community that forms a genetic unit. See Dobzhansky (1950); Mayr (1969); Simpson (1943). *Synonyms*: Gentes (singular: Gens).

14. Genotypic cluster

Clusters of monotypic or polytypic biological entities, identified using morphology or genetics, forming groups that have few or no intermediates when in contact. See Mallet (1995). *Synonyms*: Polythetic species.

15. Hennigian species

A phylospecies conception. A tokogenetic community that arises when a stem species is dissolved into two new species and ends when it goes extinct or speciates. See Hennig (1966; 1950); Meier and Willman (1997). *Synonyms*: Biospecies (in part), cladospecies (in part), phylospecies (in part), internodal species.

16. Internodal species

Organisms are conspecific in virtue of their common membership of a part of a genealogical network between two permanent splitting events or a splitting event and extinction. See Kornet (1993). *Synonyms*: Cladospecies and Hennigian species (in part), phylospecies.

17. Least Inclusive Taxonomic Unit (LITUs)

A taxonomic group that is diagnosable in terms of its autapomorphies, but has no fixed rank or binomial. (Pleijel 1999; Pleijel and Rouse 2000). *Synonyms*: Deme (Gilmour 1958; cf. Winsor 2000).

18. Morphospecies*

Defined by Aristotle and Linnaeus, and too many others to name, but including Owen, Agassiz, and recently, Cronquist. Species are the smallest groups that are consistently and persistently distinct, and distinguishable by ordinary means. Contrary to the received view, this was never anything more than a diagnostic account of species. See Cronquist (1978). *Synonyms*: Classical species, Linnaean species.

19. Non-dimensional species

Species delimitation in a non-dimensional system (a system without the dimensions of space and time). See Mayr (1942; 1963). *Synonyms*: Folk taxonomical kinds (Atran 1990).

20. Nothospecies

Species formed from the hybridization of two distinct parental species, often by polyploidy. See Wagner (1983). *Synonyms*: hybrid species, reticulate species.

21. Phylogenetic Taxon species

A phylospecies conception: A species is the smallest diagnosable cluster of individual organisms within which there is a parental pattern of ancestry and descent, Cracraft (1983); Eldredge and Cracraft (1980); the least inclusive taxon recognized in a classification, into which organism are grouped because of evidence of monophyly (usually, but not restricted to, the presence of synapomorphies), that is ranked as a species because it is the smallest 'important' lineage deemed worthy of formal recognition, where 'important' refers to the action of those processes that are dominant in producing and maintaining lineages in a particular case, Nixon and Wheeler (1990). See Mishler and Brandon (1987).

22. Phenospecies

A cluster of characters that statistically covary, a family resemblance concept in which possession of most characters is required for inclusion in a species, but not all. A class of organisms that share most of a set of characters. See Beckner (1959); Sokal and Sneath (1963). *Synonyms*: Operational Taxonomic Unit (OTU), Phena (singular phenon) (Smith 1994).

23. Recognition species

Specifications: A species is that most inclusive population of individual, biparental organisms which share a common fertilization system. See Paterson (1985). *Synonyms*: Specific mate recognition system (SMRS).

24. Reproductive competition species

The most extensive units in the natural economy such that reproductive competition occurs among their parts. See Ghiselin (1974b). *Synonyms*: Hypermodern species concept.

25. Successional species

Arbitrary anagenetic stages in morphological forms, mainly in the paleontological record. See George (1956); Simpson (1961). *Synonyms*: Paleospecies, evolutionary species (in part), chronospecies.

26. Taxonomic species*

Specimens considered by a taxonomist to be members of a kind on the evidence or on the assumption they are as alike as their offspring of hereditary relatives within a few generations. Whatever a competent taxonomist chooses to call a species. See Blackwelder (1967), but see also Regan (1926) and Strickland *et al.*(1843). *Synonyms*: Cynical species concept (Kitcher 1984).

⚓ Bibliography

Adanson, Michel. 1763. Familles des plantes: I. Partie. Contenant une Préface Istorike sur l'état ancien & actuel de la Botanike, & une Téorie de cete Science. Paris: Vincent.

Agassiz, Louis. 1842. New views regarding the distribution of fossils in formations. *Edinburgh New Philosophical Journal* 32 (63):97-98.

———. 1859. *An essay on classification*. London: Longman, Brown, Green, Longmans and Roberts and Trubner.

———. 1863. *Methods of study in natural history*. Boston: Ticknor and Fields.

———. 1869. *De l'Espece et de la Classification en Zoologie*. Paris: Balliere.

Aguilar, Javier Fuertes, Josep Antoni Roselló, and Gonzalo Nieto Feliner. 1999. Molecular evidence for the compilospecies model of reticulate evolution in *Armeria* (Plumbaginaceae). *Systematic Biology* 48 (4):735-754.

Albertus Magnus. 1987. *Man and the beasts: De animalibus (Books 22-26)*. Translated by J. J. Scanlan. Binghampton, N.Y.: Medieval & Renaissance Texts & Studies.

Allen, Garland E. 1980. The evolutionary synthesis: Morgan and natural selection revisited. In *The evolutionary synthesis*, edited by E. Mayr and W. B. Provine. New York: Columbia University Press.

Amundson, Ron. 2005. *The changing rule of the embryo in evolutionary biology: structure and synthesis, Cambridge studies in philosophy and biology*. New York: Cambridge University Press.

Anderson, Lorin. 1976. Charles Bonnet's taxonomy and the Chain of Being. *Journal of the History of Ideas* 37 (1):45-58.

Appel, Toby A. 1987. *The Cuvier-Geoffroy debate: French biology in the decades before Darwin, Monographs on the history and philosophy of biology*. New York: Oxford University Press.

Aquinas, Thomas. 1947. *Summa theologica*. 1st complete American ed. New York,: Benziger Bros.

Argyll, George J. D. Campbell, the Duke of. 1884. *The reign of law*. 18th ed. London: Alexander Strahan. Original edition, 1866.

Aristotle. 1995. *The complete works of Aristotle: the revised Oxford translation*. Edited by J. Barnes, Bollingen series ; 71:2. Princeton, N.J.: Princeton University Press.

———. 1998. *The Metaphysics*. Translated by H. Lawson-Tancred. London: Penguin.

Atran, Scott. 1985. The early history of the species concept: an anthropological reading. In *Histoire du Concept D'Espece dans les Sciences de la Vie*. Paris: Fondation Singer-Polignac.

———. 1990. *The cognitive foundations of natural history*. New York: Cambridge University Press.

———. 1995. Causal constraints on categories and categorical constraints on biological reasoning across cultures. In *Causal cognition: a multidisciplinary debate*, edited by D. Sperber, D. Premack and A. J. Premack. Oxford, UK: New York: Clarendon Press; Oxford University Press.

———. 1998. Folk biology and the anthropology of science: cognitive universals and the cultural particulars. *Behavioral and Brain Sciences* 21 (4):547–609.

———. 1999. The universal primacy of generic species in folkbiological taxonomy: Implications for human biological, cultural and scientific evolution. In *Species, New interdisciplinary essays*, edited by R. A. Wilson. Cambridge, MA: Bradford/MIT Press.

Augustine, Saint Bishop of Hippo. 1982. *The literal meaning of Genesis*. Translated by J. H. Taylor, *Ancient Christian writers; no. 41–42*. New York, N.Y.: Newman Press.

Avise, J. C., and R. M. Ball Jr. 1990. Principles of genealogical concordance in species concepts and biological taxonomy. In *Oxford Surveys in Evolutionary Biology*, edited by D. Futuyma and J. Atonovics. Oxford: Oxford University Press.

Babbage, Charles. 1837. *The Ninth Bridgewater Treatise: A fragment*. London: John Murray.

Bacon, Francis. 1863. *The Works of Francis Bacon*. Edited by J. Spedding, R. L. Ellis and D. D. Heath. Vol. II: Sylva Sylvarum. Boston: Taggard and Thompson.

———. 1960. *The new Organon and related writings*. Translated by F. H. Anderson. Indianapolis: Bobbs-Merrill. Original edition, 1620.

Balme, D. M. 1987a. Aristotle's biology was not essentialist. In *Philosophical issues in Aristotle's biology*, edited by A. Gotthelf and J. G. Lennox. Cambridge UK: Cambridge University Press.

———. 1987b. The place of biology in Aristotle's philosophy. In *Philosophical issues in Aristotle's biology*, edited by A. Gotthelf and J. G. Lennox. Cambridge UK: Cambridge University Press.

Barlow, Nora, ed. 1967. *Darwin and Henslow: the growth of an idea; letters, 1831–1860*. London: Murray [for] Bentham-Moxon Trust.

Barnes, J., ed. 1984. *The complete works of Aristotle*. 2 vols. Vol. 1. Princeton NJ: Princeton University Press.

Barthélemy-Madaule, Madeleine. 1982. *Lamarck, the mythical precursor: a study of the relations between science and ideology*. Cambridge, Mass.: MIT Press.

Basalla, George, William Coleman, and Robert H. Kargon, eds. 1970. *Victorian science: a self-portrait from the presidential addresses of the British Association for the Advancement of Science*. New York: Anchor/Doubleday.

Bateson, William. 1894. *Material for the study of variation treated with especial regard to discontinuity in the origin of species*. London: Macmillan.

Beatty, J. 1985. Speaking of species: Darwin's strategy. In *The Darwinian heritage*, edited by D. Kohn. Princeton: Princeton University Press.

Beckner, M. 1959. *The biological way of thought*. New York: Columbia University Press.

Beltran, M., C. D. Jiggins, V. Bull, M. Linares, J. Mallet, W. O. McMillan, and E. Bermingham. 2002. Phylogenetic discordance at the species boundary: comparative gene genealogies among rapidly radiating heliconius butterflies. *Mol Biol Evol* 19 (12):2176–90.

Bentham, George. 1827. *An Outline of a New System of Logic. With a Critical Examination of Dr. Whately's "Elements of Logic"*. London: Hunt and Clark.

Bessey, Charles E. 1908. The Taxonomic Aspect of the Species Question. *The American Naturalist* 42 (496):218–224.

Bigelow, R. S. 1965. Hybrid zones and reproductive isolation. *Evolution* 19 (4):449–458.

———. 1967. *The grasshoppers (Acrididae) of New Zealand: their taxonomy and distribution*. Christchurch: University of Canterbury.

Blackwelder, Richard E. 1967. *Taxonomy: a text and reference book*. New York: Wiley.

Boodin, John Elof. 1943. The discovery of form. *Journal of the History of Ideas* 4 (2):177–192.

Britton, Nathaniel Lord. 1908. The taxonomic aspect of the species question. *American Naturalist* 42 (496):225–242.

Brower, Andrew V. Z. 2002. Cladistics, populations and species in geographical space: the case of *Heliconius* butterflies. In *Molecular systematics and evolution: Theory and practice*, edited by R. DeSalle, G. Girbet and W. Wheeler. Switzerland: Birkhäuser Verlag.

Burg, Theresa M., Andrew W. Trites, and Michael J. Smith. 1999. Mitochondrial and microsatellite DNA analyses of harbour seal population structure in the northeast Pacific Ocean. *Can. J. Zool.* 77:930–943.

Burkhardt, Frederick, ed. 1996. *Charles Darwin's letters: a selection, 1825–1859*. Cambridge; New York, NY, USA: University of Cambridge.

Burkhardt, Richard W. 1985. Lamarck and species. In *Histoire du Concept D'Espece dans les Sciences de la Vie*. Paris: Fondation Singer-Polignac.

Burnet, Thomas. 1681 [1684–1689 English edition]. *Telluris Theoria Sacra, or Sacred Theory of the Earth*. London: R. Norton, for W. Kettilby.

Butler, Samuel. 1879. *Evolution, old and new, or, The theories of Buffon, Dr. Erasmus Darwin, and Lamarck, as compared with that of Mr. Charles Darwin*. London: Hardwicke and Bogue.

Cain, A. J. 1995. Linnaeus's natural and artificial arrangements of plants. *Botanical Journal of the Linnean Society* 117 (2):73.

Cain, Arthur J. 1954. *Animal species and their evolution*. London: Hutchinson University Library.

———. 1994. Numerus, figura, proportio, situs: Linnaeus's definitory attributes. *Archives of Natural History* 21:17–36.

———. 1999. Thomas Sydenham, John Ray, and some contemporaries on species. *Archives of Natural History* 24 (1):55–83.

Camardi, Giovanni. 2001. Richard Owen, Morphology and Evolution. *Journal of the History of Biology* 34 (3):481.

Camp, W. H. 1951. Biosystematy. *Brittonia* 7:113–127.

Camp, W. H., and C. L. Gilly. 1943. The structure and origin of species. *Brittonia* 4:325–385.

Candolle, Augustin-Pyramus, and Kurt Sprengel. 1821. *Elements of the Philosophy of Plants*. Edinburgh: W. Blackwood.

Candolle, Augustine-Pyramus de. 1819. *Théorie élementaire de la botanique, ou exposition des principes de la classification naturelle et de l'art de décrire et d'étudier les végétaux*. 2nd ed. Paris.

Cantino, Philip D., and Kevin de Querioz. *Phylocode: A phylogenetic code of biological nomenclature* 2000 [cited 23 March 2001. Available from http://www.ohiou.edu/phylocode/.

Cassirer, Ernst, Paul Oskar Kristeller, and John Herman Randall, eds. 1948. *The Renaissance philosophy of man: selections in translation*. Chicago: University of Chicago Press.

Chambers, Robert. 1844. *Vestiges of the natural history of creation*. London: John Churchill.

Chung, Carl. 2003. On the origin of the typological/population distinction in Ernst Mayr's changing views of species, 1942–1959. *Studies in History and Philosophy of Biological and Biomedical Sciences* 34:277–296.

Clarke, Richard F. 1895. *Logic*. 3rd ed, *Manuals of Catholic Philosophy*. London: Longmans, Green.

Coleman, William R. 1964. *Georges Cuvier, zoologist: a study in the history of evolution theory*. Cambridge, Mass.: Harvard University Press.

Coyne, Jerry A., and H. Allen Orr. 2004. *Speciation*. Sunderland, Mass.: Sinauer Associates.

Cracraft, Joel. 1983. Species concepts and speciation analysis. In *Current Ornithology*, edited by R. F. Johnston. New York: Plenum Press.

———. 1997. Species concepts in systematics and conservation biology—an ornithological viewpoint. In *Species: The units of biodiversity*, edited by M. F. Claridge, H. A. Dawah and M. R. Wilson. London: Chapman and Hall.

———. 2000. Species concepts in theoretical and applied biology: A systematic debate with consequences. In *Species concepts and phylogenetic theory: A debate*, edited by Q. D. Wheeler and R. Meier. New York: Columbia University Press.

Cronquist, A. 1978. Once again, what is a species? In *BioSystematics in Agriculture*, edited by L. Knutson. Montclair, NJ: Alleheld Osmun.

Cuvier, Georges. 1812. *Discours sur les révolutions du globe (Discourse on the Revolutionary Upheavals on the Surface of the Earth)*. Translated by I. Johnston. Paris.

Darlington, Cyril Dean. 1940. Taxonomic species and genetic systems. In *The new systematics*, edited by J. Huxley. London: Oxford University Press.

Darwin, Charles. 1839. *Journal of researches into the geology and natural history of the various countries visited by H.M.S. Beagle under the command of Captain Fitzroy, R.N. from 1832 to 1836*. London: Henry Colburn.

———. 1851. *A monograph on the sub-class Cirripedia: with figures of all the species*. London: Printed for the Ray Society.

———. 1868. *The variation of animals and plants under domestication*. London: John Murray.

———. 1869. *On the origin of species by means of natural selection, or the preservation of favoured races in the struggle for life*. 5th ed. London: John Murray.

———. 1871. *The descent of man and selection in relation to sex*. London: John Murray.

———. 1873. Origin of certain instincts. *Nature* vii:417.

———. 1875. *The variation of animals and plants under domestication*. 2nd revised ed. London: John Murray. Original edition, 1868.

———. 1888a. *The life and letters of Charles Darwin: including an autobiographical chapter*. Edited by F.Darwin. Lond.: Murray.

———. 1998. *The variation of animals and plants under domestication*. 2nd ed. 2 vols. Baltimore: Johns Hopkins University Press. Original edition, 1888 Appleton American edition.

Darwin, Francis, ed. 1888b. *The life and letters of Charles Darwin: including an autobiographical chapter*. 3 vols. London: John Murray.

Davis, Jerrold I. 1997. Evolution, evidence, and the role of species concepts in phylogenetics. *Systematic Biology* 22 (2):373–403.

de Beer, Gavin Rylands. 1960. Darwin's notebooks on transmutation of species. Part I. First notebook [B] (July 1837-February 1838). *Bulletin of the British Museum (Natural History)* Historical Series 2 (2): 23-73.

de Queiroz, Kevin. 1998. The general lineage concept of species, species criteria, and the process of speciation. In *Endless forms: species and speciation*, edited by D. J. Howard and S. H. Berlocher. New York: Oxford University Press.

———. 1999. The general lineage concept of species and the defining properties of the species category. In *Species, New interdisciplinary essays*, edited by R. A. Wilson. Cambridge, MA: Bradford/MIT Press.

de Queiroz, Kevin, and Michael J Donoghue. 1988. Phylogenetic systematics and the species problem. *Cladistics* 4:317–338.

———. 1990. Phylogenetic systematics and species revisited. *Cladistics* 6:83–90.

De Vries, Hugo. 1912. *Species and Varieties: their origin by mutation. Lectures delivered at the University of California*. 3rd ed. Chicago: Open Court. Original edition, 1904.

Dewey, John. 1997. *The influence of Darwin on philosophy and other essays, Great books in philosophy*. Amherst, N.Y.: Prometheus Books.

Dobzhansky, Theodosius. 1935. A critique of the species concept in biology. *Philosophy of Science* 2:344–355.

———. 1937. *Genetics and the origin of species*. New York: Columbia University Press.

———. 1941. *Genetics and the origin of species*. 2nd ed. New York: Columbia University Press.

———. 1950. Mendelian populations and their evolution. *American Naturalist* 74:312–321.

———. 1962. *Mankind evolving; the evolution of the human species*. New Haven: Yale University Press.

———. 1970. *Genetics of the evolutionary process*. New York: Columbia University Press.

Dres, M., and J. Mallet. 2002. Host races in plant-feeding insects and their importance in sympatric speciation. *Philos Trans R Soc Lond B Biol Sci* 357 (1420):471–92.

Eigen, Edward A. 1997. Overcoming First Impressions: Georges Cuvier's Types. *Journal of the History of Biology* 30 (2):179.

Eigen, Manfred. 1993. Viral quasispecies. *Scientific American* July 1993 (32–39).

Eldredge, Niles. 1989. *Macroevolutionary dynamics: species, niches, and adaptive peaks*. New York: McGraw-Hill.

———. 1993. What, if anything, is a species? In *Species, species concepts, and primate evolution*, edited by W. H. Kimbel and L. B. Martin. New York: Plenum Press.

Eldredge, Niles, and Joel Cracraft. 1980. *Phylogenetic patterns and the evolutionary process: method and theory in comparative biology*. New York: Columbia University Press.

Ereshefsky, Marc. 1999. Species and the Linnean hierarchy. In *Species, New interdisciplinary essays*, edited by R. A. Wilson. Cambridge, MA: Bradford/MIT Press.

———. 2000. *The poverty of Linnaean hierarchy: a philosophical study of biological taxonomy*. Cambridge, UK; New York: Cambridge University Press.

Euzéby:, J.P. 2006. *List of Prokaryotic Names with Standing in Nomenclature* 2006 [cited 17/2/2006 2006]. Available from http://www.bacterio.cict.fr/.

Farber, Paul. 1976. The type-concept in zoology during the first half of the nineteenth century. *Journal of the History of Biology* 9 (1):93–119.

Farber, Paul Lawrence. 1971. Buffon's concept of species. PhD, Bloomington,.

Fisher, RA. 1930. *The genetical theory of natural selection*. Oxford UK: Clarendon Press, (rev. ed. Dover, New York, 1958).

Forsdyke, Donald R. 2001. *The origin of species revisited: a Victorian who anticipated modern developments in Darwin's theory*. Kingston, Ont.: McGill-Queen's University Press.

Foucault, Michel. 1970. *The order of things: an archaeology of the human sciences*. London: Routledge Classics.

Futuyma, Douglas J. 1983. *Science on trial: The case for evolution*. New York: Pantheon.

Gadow, Hans. 1909. XVII. Geographical Distribution of Animals:. In *Darwin and modern science; essays in commemoration of the centenary of the birth of Charles Darwin and of the fiftieth anniversary of the publication of the Origin of species*, edited by A. C. Seward. Cambridge: Cambridge University press.

Gasking, Elizabeth B. 1967. *Investigations into generation 1651–1828, History of scientific ideas*. London: Hutchinson.

Geoffroy Saint Hilaire, Isidore. 1859. *Histoire naturelle générale des règnes organiques, principalement étudiée chez l' homme et les animaux*. Vol. II. Paris: Victor Masson.

George, T. N. 1956. Biospecies, chronospecies and morphospecies. In *The species concept in paleontology*, edited by P. C. Sylvester-Bradley. London: Systematics Association.

Ghiselin, Michael T. 1966. On psychologism in the logic of taxonomic controversies. *Systematic Zoology* 15:207–215.

———. 1974a. A radical solution to the species problem. *Systematic Zoology* 23:536–544.

———. 1974b. *The economy of nature and the evolution of sex*. Berkeley: University of California Press.

———. 1984. *The triumph of the Darwinian method, with a new preface*. rev. ed. Chicago: University of Chicago Press. Original edition, 1969.

———. 1997. *Metaphysics and the origin of species*. Albany: State University of New York Press.

———. 2005. The Darwinian Revolution as Viewed by a Philosophical Biologist. *Journal of the History of Biology* 38 (1):123.

Gilmour, J. S. L. 1940. Taxonomy and philosophy. In *The new systematics*, edited by J. Huxley. London: Oxford University Press.

———. 1958. The Species: Yesterday and To-Morrow. *Nature* 181 (4606):379–380.

Glass, Bentley. 1959. The germination of the idea of biological species. In *Forerunners of Darwin, 1745-1859*, edited by B. Glass, O. Temkin and W. L. Straus Jr. Baltimore: Johns Hopkins Press.

Goethe, Johann Wolfgang von, Georg Christoph Tobler, and Agnes Arber. 1946. *Goethe's botany: the metamorphosis of plants (1790) and Tobler's Ode to nature (1782)*. [Waltham? Mass.: Chronica Botanica.

Goldschmidt, Richard B. 1940. *The material basis of evolution*. Seattle: University of Washington Press.

Gosse, Edmund. 1970. *Father and Son: A study of two temperaments*. London: Heinemann. Original edition, Heinemann, 1907.

Gosse, Philip Henry. 1857. *Creation (Omphalos): an attempt to untie the geological knot*. London: J. Van Voorst.

Gould, Steven Jay. 1977. The Reverend Thomas' Dirty Little Planet. In *Ever Since Darwin*. N.Y.: W.W. Norton & Co.

Grant, Verne. 1957. The Plant Species in Theory and Practice. In *The Species Problem*, edited by E. Mayr. Washington DC: American Association for the Advancement of Science.

Gray, Asa. 1879. *Structural botany, or Organography on the basis of morphology. To which is added the principles of taxonomy and phytography, and a glossary of botanical terms*. New York, Chicago: Ivison, Blakeman, Taylor.

Greene, John C. 1959. *The death of Adam: evolution and its impact on Western thought*. Ames: Iowa State University Press.

———. 1963. *Darwin and the modern world view: The Rockwell Lectures, Rice University*. New York: New American Library. Original edition, 1961.

Gregg, John Richard. 1954. *The language of taxonomy: an application of symbolic logic to the study of classificatory systems*. New York: Columbia University Press.

Grew, Nehemiah. 1682. *The anatomy of plants with an idea of a philosophical history of plants, and several other lectures, read before the royal society, Early English Books, 1641-1700/456:17*: London: Printed by W. Rawlins ...

Gustafsson, Ö. 1979. Linnaeus' Peloria: The history of a monster. *TAG Theoretical and Applied Genetics* 54 (6):241–248.

Haeckel, Ernst Heinrich Philipp August. 1883. *The evolution of man: a popular exposition of the principal points of human ontogeny and phylogeny*. 2 vols. London: Kegan, Paul, Trench & Co.

Haldane, J. B. S. 1956. Can a species concept be justified? In *The species concept in palaeontology: A symposium*, edited by P. C. Sylvester-Bradley. London: The Systematics Association.

Harlan, J. R., and J. M. J. De Wet. 1963. The compilospecies concept. *Evolution* 17:497–501.

Hennig, Willi. 1950. *Grundzeuge einer Theorie der Phylogenetischen Systematik*. Berlin: Aufbau Verlag.

———. 1966. *Phylogenetic systematics*. Translated by D. D. Davis and R. Zangerl. Urbana: University of Illinois Press.

Hey, Jody. 2001. *Genes, concepts and species: the evolutionary and cognitive causes of the species problem*. New York: Oxford University Press.

Hooke, Robert. 1665. *Micrographia: or some Physiological Descriptions of Minute Bodies made by Magnifying Glasses with Observations and Inquiries Thereupon*. London: Jo. Martyn, and Ja. Allestry.

Hopkins, Jasper. 1981. *Nicholas of Cusa On learned ignorance: a translation and an appraisal of De docta ignorantia*. Minneapolis: A.J. Benning Press.

Hull, David L. 1965. The effect of essentialism on taxonomy: Two thousand years of stasis. *British Journal for the Philosophy of Science* 15:314–326, 16:1–18.

———. 1976. Are species really individuals? *Systematic Zoology* 25:174–191.

———. 1981. Units of evolution: a metaphysical essay. In *The philosophy of evolution*, edited by U. L. Jensen and R. Harré. Brighton UK: Harvester Press.

———. 1984a. Historical entities and historical narratives. In *Minds, machines, and evolution*, edited by C. Hookway. Cambridge: Cambridge University Press.

———. 1984b. Lamarck among the Anglos. In *Introduction to reprinted edition of J. B. Lamarck's Zoological Philosophy: An Exposition with Regard to the Natural History of Animals*. Chicago: Chicago University Press.

———. 1988. *Science as a process: an evolutionary account of the social and conceptual development of science*. Chicago: University of Chicago Press.

———. 1999. On the plurality of species: Questioning the party line. In *Species, New interdisciplinary essays*, edited by R. A. Wilson. Cambridge, MA: Bradford/MIT Press.

———, ed. 1973. *Darwin and his critics; the reception of Darwin's theory of evolution by the scientific community*. Cambridge, Mass.: Harvard University Press.

Hull, David L., and Michael Ruse, eds. 1998. *The philosophy of biology*. Oxford; New York: Oxford University Press.

Hunter Dupree, A. 1968. *Asa Gray 1810–1888*. College ed. Vol. 132. New York: Atheneum. Original edition, Belknap 1959.

Hunter, John. 1787. Observations Tending to Shew That the Wolf, Jackal, and Dog, are All of the Same Species. *Philosophical Transactions of the Royal Society of London Series B-Biological Sciences* 77:253–266.

Huxley, Julian. 1942. *Evolution: the modern synthesis*. London: Allen and Unwin.

Huxley, T. H. 1906. *Man's place in nature and other essays*. Everyman's Library ed. London; New York: J. M. Dent/E. P. Dutton.

Huxley, Thomas Henry. 1893. *Darwiniana: essays*. London Macmillan.

Jacot, Arthur Paul. 1932. The Status of the Species and the Genus. *American Naturalist* 66 (705):346–364.

Jevons, William Stanley. 1878. *The principles of science: a treatise on logic and scientific method*. 2nd ed. London: Macmillan. Original edition, 1873.

Jordan, Karl. 1905. Der Gegensatz zwischen geographischer und nichtgeographischer Variation. *Zeitschrift für Wissenschaftliche Zoologie* 83:151–210.

Jordanova, L. J. 1984. *Lamarck, Past masters*. Oxford; New York: Oxford University Press.

Joseph, H. W. B. 1916. *An introduction to logic*. 2nd ed. Oxford: Clarendon Press.

Jussieu, Antoine-Laurent de. 1964. *Genera plantarum*. Facsimile ed. Weinheim, Germany; Codicote, UK; New York: J. Cramer. Original edition, 1789, Paris.

Kant, Immanuel. 1933. *Critique of pure reason*. Translated by N. K. Smith. 2nd revised ed. London: Macmillan. Original edition, 1787 second edition.

———. 1951. *Critique of judgment*. Translated by J. H. Bernard. New York: Hafner. Original edition, 1790/1793.

Kitcher, Philip. 1984. Species. *Philosophy of Science* 51:308–333.

Koerner, Lisbet. 1999. *Linnaeus: nature and nation*. Cambridge, Mass: Harvard University Press.

Kornet, D. 1993. Internodal species concept. *J Theor Biol* 104:407–435.

Kornet, D, and JW McAllister. 1993. The composite species concept. In *Reconstructing species: Demarcations in genealogical networks*. Rijksherbarium, Leiden: Unpublished phD dissertation, Institute for Theoretical Biology.

Kottler, Malcolm J. 1978. Charles Darwin's biological species concept and theory of geographic speciation: the Transmutation Notebooks. *Annals of Science* 35:275–297.

Kvist, L., J. Martens, H. Higuchi, A. A. Nazarenko, O. P. Valchuk, and M. Orell. 2003. Evolution and genetic structure of the great tit (Parus major) complex. *Proc R Soc Lond B Biol Sci* 270 (1523):1447–54.

Lamarck, Jean Baptiste. 1809. *Philosophie zoologique, ou, Exposition des considérations relative à l'histoire naturelle des animaux*. Paris: Dentu.

———. 1914. *Zoological philosophy: an exposition with regard to the natural history of animals*. Translated by H. Elliot. London: Macmillan.

Lambert, David M., and Hamish G. Spencer, eds. 1995. *Speciation and the recognition concept: theory and application*. Baltimore: Johns Hopkins University Press.

Lankester, Edwin Ray. 1890. *The advancement of science. Occasional essays & addresses*. London and New York: Macmillan and Co.

Larson, James L. 1967. Linnaeus and the Natural Method. *Isis* 58 (3):304–320.

———. 1968. The Species Concept of Linnaeus. *Isis* 59 (3):291–299.

Leibniz, Gottfried Wilhelm. 1996. *New Essays on Human Understanding*. Translated by P. Remnant and J. Bennett. Cambridge UK: Cambridge University Press. Original edition, 1765.

Lennox, James G. 1987. Kinds, forms of kinds, and the more and the less in Aristotle's biology. In *Philosophical issues in Aristotle's biology*, edited by A. Gotthelf and J. G. Lennox. Cambridge UK: Cambridge University Press.

Lennox, James G. 1993. Darwin was a teleologist. *Biology and Philosophy* 8 (4):409–421.

———. 1994. Aristotle's biology: plain, but not simple. *Stud Hist Philos Sci* 25 (5):817–23.

———. 2001. *Aristotle's philosophy of biology: studies in the origins of life science*. Cambridge, UK; New York: Cambridge University Press.

Lenoir, Timothy. 1987. The eternal laws of form: Morphotypes and the conditions of existence in Goethe's biological thought. In *Goethe and the Sciences: A Re-appraisal* edited by F. Amrine, F. J. Zucker and H. Wheeler. Berlin; New York: Springer Verlag.

Levit, Georgy S., and Kay Meister. 2006. The history of essentialism vs. Ernst Mayr's "Essentialism Story": A case study of German idealistic morphology *Theory in Biosciences* 124:281–307.

Lewes, George Henry. 1860. *The Physiology of Common Life*. 2 vols. Vol. 2. Edinburgh: W. Blackwood.

Lewes, George Henry, [Anon.]. 1856. Hereditary Influence, Animal and Human. *Westminster Review* 66 (July):135–162.

Lherminer, Philippe, and Michel Solignac. 2000. L'espèce: définitions d'auters. *Sciences de la vie*:153–165.

Liddell, H. G., and Scott. 1888. *An intermediate Greek–English lexicon, founded upon the seventh edition of Liddell and Scott's Greek–English Lexicon*. Oxford: Clarendon Press.

Liebers, Dorit, Peter de Knijff, and Andreas J. Helbig. 2004. The herring gull complex is not a ring species *Proc. R. Soc. Lond. B* 271:893–901.

Lindley, John. 1832. *An Introduction to Biology*. London: Longman, Orme, Brown, Green, and Longmans.

———. 1841. *Elements of Botany, Structural, Physiological, Systematical, and Medical*. 4th ed. London: Taylor and Walton.

Linne, Carl von. 1788-1793. *Systema naturae per regna tria naturae, secundum classes, ordines, genera, species, cum characteribus, differentiis, synonymis, locis ... /cura Jo. Frid. Gmelin*. Editio decima tertia, aucta, reformata. ed. Lipsiae: Impensis Georg. Emanuel Beer.

Lotsy, J. P. 1916. *Evolution by means of hybridization*. The Hague: Martinus Nijhoff.

Lovejoy, Arthur O. 1936. *The great chain of being: a study of the history of an idea*. Cambridge, Mass.: Harvard University Press.

———. 1946. Goldsmith and the Chain of Being. *Journal of the History of Ideas* 7 (1):91-98.

———. 1959. Buffon and the problem of species. In *Forerunners of Darwin 1745-1859*, edited by B. Glass, O. Temkin and W. L. Straus. Baltimore, MD: Johns Hopkins Press.

Lucretius. 1969. *On the nature of things (De rerum natura)*. Translated by M. F. Smith. London: Sphere Books.

Lurie, Edward. 1960. *Louis Agassiz: A life in science*. Baltimore and London: Johns Hopkins University Press.

Lustig, A. J. 2002. Erich Wasmann, Ernst Haeckel, and the Limits of Science. *Theory in Biosciences* 121 (3):252-259.

Lyell, Charles. 1832. *Principles of geology, being an attempt to explain the former changes of the earth's surface, by reference to causes now in operation*. 2nd ed. London: John Murray.

Mallet, J. 1995. The species definition for the modern synthesis. *Trends in Ecology and Evolution* 10 (7):294-299.

Mallet, James. 2000. Species and their names. *Trends in Ecology and Evolution* 15 (8):344-345.

———. 2001. Species, concepts of. In *Encyclopedia of biodiversity*, edited by S. A. Levin. New York: Academic Press.

Mandelbaum, Maurice. 1957. The Scientific Background of Evolutionary Theory in Biology. *Journal of the History of Ideas* 18 (3):342-361.

Mayden, R. L. 1997. A hierarchy of species concepts: the denoument in the saga of the species problem. In *Species: The units of diversity*, edited by M. F. Claridge, H. A. Dawah and M. R. Wilson. London: Chapman and Hall.

———. 2002. On biological species, species concepts and individuation in the natural world. *Fish and Fisheries* 3 (3):171-196.

Mayr, Ernst. 1940. Speciation phenomena in birds. *American Naturalist* 74:249-278.

———. 1942. *Systematics and the origin of species from the viewpoint of a zoologist*. New York: Columbia University Press.

———. 1957. Species concepts and definitions. In *The species problem: A symposium presented at the Atlanta meeting of the American Association for the Advancement of Science, December 28-29, 1955, Publication No 50*, edited by E. Mayr. Washington DC: American Association for the Advancement of Science.

———. 1963. *Animal species and evolution*. Cambridge MA: The Belknap Press of Harvard University Press.

———. 1969. *Principles of systematic zoology*. New York: McGraw-Hill.

———. 1970. *Populations, species, and evolution: an abridgment of Animal species and evolution*. Cambridge, Mass.: Belknap Press of Harvard University Press.

———. 1982. *The growth of biological thought: diversity, evolution, and inheritance*. Cambridge, Mass.: Belknap Press.

———. 1991. *One long argument: Charles Darwin and the genesis of modern evolutionary thought*. Cambridge, Mass.: Harvard University Press.

———. 1995. Species, classification, and evolution. In *Biodiversity and evolution*, edited by R. Arai, M. Kato and Y. Doi. Tokyo: National Science Museum Foundation.

———. 1996. What is a species, and what is not? *Philosophy of Science* 2:262–277.

———. 1999. *Systematics and the origin of species from the viewpoint of a zoologist*. New York: Columbia University Press. Original edition, 1942.

———. 2000a. A critique from the biological species concept: what is a species, and what is not? In *Species concepts and phylogenetic theory: A debate*, edited by Q. D. Wheeler and R. Meier. New York: Columbia University Press.

———. 2000b. The biological species concept. In *Species concepts and phylogenetic theory: A debate*, edited by Q. D. Wheeler and R. Meier. New York: Columbia University Press.

Mayr, Ernst, and Peter D. Ashlock. 1991. *Principles of systematic zoology*. 2nd ed. New York: McGraw-Hill,.

Mayr, Ernst, and William B. Provine. 1980. *The Evolutionary synthesis : perspectives on the unification of biology*. Cambridge, Mass: Harvard University Press.

McKeon, Richard. 1929. *Selections from medieval philosophers*. 2 vols. Vol. 1. New York: Charles Scribners Sons.

———. 1930. *Selections from medieval philosophers*. 2 vols. Vol. 2. New York: Charles Scribners Sons.

———, ed. 1941. *The basic works of Aristotle*. New York: Random House.

McOuat, Gordon. 2003. The logical systematist: George Bentham and his *Outline of a new system of logic*. *Archives of Natural History* 30 (2):203–223.

Meier, Rudolf, and Rainer Willmann. 1997. The Hennigian species concept. In *Species concepts and phylogenetic theory: A debate*, edited by Q. Wheeler and R. Meier. New York: Columbia University Press.

———. 2000. The Hennigian species concept. In *Species concepts and phylogenetic theory: A debate*, edited by Q. Wheeler and R. Meier. New York: Columbia University Press.

Mendel, J. Gregor. 1866. Versuche über Plflanzenhybriden. *Verhandlungen des naturforschenden Vereines in Brünn* Bd. IV für das Jahr, 1865 Abhandlungen (3–47).

Mill, John Stuart. 1974. *A System of Logic, Ratiocinative and Inducfve, Being a Connected View of the Principles of Evidence and the Methods of Scientific Investigation*. Edited by J. M. Robson. Vol. 8, *The Collected Works of John Stuart Mill*. Toronto; Buffalo: University of Toronto Press/ Routledge & Kegan Paul.

Mishler, Brent D., and Robert N. Brandon. 1987. Individuality, pluralism, and the Phylogenetic Species Concept. *Biology and Philosophy* 2:397–414.

Mishler, Brent D., and Michael J. Donoghue. 1982. Species concepts: A case for pluralism. *Systematic Zoology* 31:491–503.

Mishler, Brent D., and Edward C. Theriot. 2000. The phylogenetic species concept (*sensu* Mishler and Theriot): Monophyly, apomorphy, and phylogenetic species concepts. In *Species concepts and phylogenetic theory: A debate*, edited by Q. D. Wheeler and R. Meier. New York: Columbia University Press.

Morgan, Thomas Hunt. 1903. *Evolution and adaptation*. New York: Macmillan.

Morris, P. J. 1997. Louis Agassiz's additions to the French translation of his *Essay on Classification*. *Journal of the History of Biology* 30:121–134.

Morton, A. G. 1981. *History of botanical science: an account of the development of botany from ancient times to the present day*. London; New York: Academic Press.

Muller, H. J. 1940. Bearings of the 'Drosophila' work on systematics. In *The new systematics*, edited by J. Huxley. London: Oxford University Press.

Müller-Wille, Staffan, and Vitezslav Orel. 2007. From Linnaean Species to Mendelian Factors: Elements of Hybridism, 1751-1870. *Annals of Science* 64 (2): 171–215.

Naisbit, R. E., C. D. Jiggins, M. Linares, C. Salazar, and J. Mallet. 2002. Hybrid Sterility, Haldane's Rule and Speciation in Heliconius cydno and H. melpomene. *Genetics* 161 (4):1517–26.

Nelson, Gareth J. 1989. Species and taxa: speciation and evolution. In *Speciation and its consequences*, edited by D. Otte and J. Endler. Sunderland, Mass.: Sinauer.

Nelson, Gareth J., and Norman I. Platnick. 1981. *Systematics and biogeography: cladistics and vicariance.* New York: Columbia University Press.

Nixon, K. C., and Q. D. Wheeler. 1990. An amplification of the phylogenetic species concept. *Cladistics* 6:211–223.

Ong, Walter J. 1958. *Ramus, method, and the decay of dialogue. From the art of discourse to the art of reason.* Cambridge MA: Harvard University Press.

Opitz, J. M. 2004. Goethe's bone and the beginnings of morphology. *Am J Med Genet* 126A (1):1–8.

Owen, Richard. 1835. On the Osteology of the Chimpanzee and Orang Utan. *Transactions of the Zoological Society* 1:343–379.

Packard, Alpheus. 1901. *Lamarck, the founder of evolution: His life and work.* New York: Longmans, Green and Co.

Padian, Kevin. 1999. Charles Darwin's view of classification in theory and practice. *Systematic Biology* 48 (2):352–364.

Paterson, Hugh E. H. 1985. The recognition concept of species. In *Species and speciation*, edited by E. Vrba. Pretoria: Transvaal Museum.

———. 1993. *Evolution and the recognition concept of species. Collected Writings.* Baltimore, MA: John Hopkins University Press.

Pellegrin, Pierre. 1986. *Aristotle's classification of animals: biology and the conceptual unity of the Aristotelian corpus.* Translated by A. Preus. Rev. ed. Berkeley: University of California Press.

———. 1987. Logical difference and biological difference: the unity of Aristotle's thought. In *Philosophical issues in Aristotle's biology*, edited by A. Gotthelf and J. G. Lennox. Cambridge UK: Cambridge University Press.

Pleijel, Frederik. 1999. Phylogenetic taxonomy, a farewell to species, and a revision of *Heteropodarke (Hesionidae, Polychaeta, Annelida)*. *Systematic Biology* 48 (4):755–789.

Pleijel, Frederik, and G. W. Rouse. 2000. Least-inclusive taxonomic unit: a new taxonomic concept for biology. *Proceedings of the Royal Society of London—Series B: Biological Sciences* 267 (1443):627–630.

Pliny, the Elder. 1855. *The Natural History.* Translated by J. Bostock and H. T. Riley. London: Taylor and Francis.

Porphyry, and Jonathan Barnes. 2003. *Porphyry's Introduction, Clarendon later ancient philosophers.* Oxford ; New York: Oxford University Press.

Porphyry, the Phoenician. 1975. *Isagoge.* Translated by E. W. Warren. Toronto: Pontifical Institute of Mediaeval Studies.

Poulton, Edward Bagnall. 1903. What is a species? *Proceedings of the Entomological Society of London*:reprinted in Poulton 1908: 46–94.

Provine, William B. 1986. *Sewall Wright and evolutionary biology, Science and its conceptual foundations.* Chicago: University of Chicago Press.

Raby, Peter. 2001. *Alfred Russel Wallace: a life.* Princeton NJ: Princeton University Press.

Ramsbottom, John. 1938. Linnaeus and the species concept. *Proceedings of the Linnean Society of London* 150 (192–220).

Regan, C. Tate. 1926. Organic evolution. *Report of the British Association for the Advancement of Science, 1925*:75–86.

Richards, Owain Westmacott, and Guy Coburn Robson. 1926. The species problem and evolution. *Nature* 117:345–347, 382–384.

Richards, Robert J. 2000. Kant and Blumenbach on the Bildungstrieb: A Historical Misunderstanding. *Stud. Hist. Phil. Biol. & Biomed. Sci* 31 (1):11–32.

———. 2005. Ernst Haeckel and the Struggles over Evolution and Religion *Annals of the History and Philosophy of Biology* 10:89–115.

Ridley, M. 1989. The cladistic solution to the species problem. *Biology and Philosophy* 4:1–16.

Robson, Guy Coburn. 1928. *The species problem: an introduction to the study of evolutionary divergence in natural populations, Biological monographs and manuals; no. 8.* Edinburgh: Oliver and Boyd.

Romanes, George J. 1886. Physiological Selection; an Additional Suggestion on the Origin of Species. *The Journal of the Linnean Society. Zoology* 19 (July 23rd):337–411.

Romanes, George John. 1895. *Darwin, and after Darwin: An exposition of the Darwinian theory and a discussion of the post-Darwinian questions.* 3 vols. Vol. II—Post Darwinian questions: heredity and utility. London: Longmans, Green and Co.

Rosen, Donn E. 1978. Vicariant patterns and historical explanation in biogeography. *Systematic Zoology* 27:159–188.

———. 1979. Fishes from the uplands and intermontane basins of Guatemala: revisionary studies and comparative biogeography. *Bulletin of the American Museum of Natural History* 162:267–376.

Ross, David. 1949. *Aristotle.* 5th ed. London: Methuen/University Paperbacks.

Rothschild, Walter, and Karl Jordan. 1906. A Revision of the American Papilios. *Novitates Zoologicae* 13 (3):411–752.

Ruse, M. 1998. All my love is toward individuals. *Evolution* 52:283–288.

Ruse, Michael. 1999. *The Darwinian revolution: science red in tooth and claw.* 2nd ed. ed. Chicago: University of Chicago Press.

Sachs, J. V. 1890. *History of Botany (1530–1860).* Translated by H. E. F. Garnsey and I. B. Balfour. Oxford: Clarendon Press. Original edition, 1875.

Sankey, Howard. 1998. Taxonomic incommensurability. *International Studies in the Philosophy of Science* 12 (1):7–16.

Shermer, Michael. 2002. *In Darwin's shadow: the life and science of Alfred Russel Wallace: a biographical study on the psychology of history.* New York: Oxford University Press.

Simpson, George Gaylord. 1943. Criteria for genera, species and subspecies in zoology and paleontology. *Annals New York Academy of Science* 44:145–178.

———. 1951. The species concept. *Evolution* 5:285–298.

———. 1961. *Principles of animal taxonomy.* New York: Columbia University Press.

Simpson, James Y. 1925. *Landmarks in the struggle between science and religion.* London: Hodder and Stoughton.

Slaughter, Mary M. 1982. *Universal languages and scientific taxonomy in the seventeenth century.* Cambridge UK; New York: Cambridge University Press.

Sloan, Phillip R. 1979. Buffon, German biology, and the historical interpretation of biological species. *British Journal for the History of Science* 12 (41):109–153.

———. 1985. From logical universals to historical individuals: Buffon's idea of biological species. In *Histoire du Concept D'Espece dans les Sciences de la Vie.* Paris: Fondation Singer-Polignac.

Smith, Andrew B. 1994. *Systematics and the fossil record: documenting evolutionary patterns.* Oxford, OX; Cambridge, Mass., USA: Blackwell Science.

Sokal, Robert R., and P. H. A. Sneath. 1963. *Principles of numerical taxonomy, A Series of books in biology.* San Francisco,: W. H. Freeman.

Sonneborn, T. M. 1957. Breeding systems, reproductive methods and species problems in Protozoa. In *The species problem: A symposium presented at the Atlanta meeting of the American Association for the Advancement of Science, December 28–29, 1955, Publication No 50*, edited by E. Mayr. Washington DC: American Association for the Advancement of Science.

Spruit, Leen. 1994–1995. *Species intelligibilis: from perception to knowledge.* 2 vols. Leiden; New York: Brill.

Stafleu, Franz Antonie. 1963. Adanson and the «Familles des plantes». In *Adanson: The bicentennial of Michel Adanson's «Familles des plantes»*, edited by G. H. M. Lawrence. Pittsburgh PA: Carnegie Institute of Technology.

Stannard, Jerry. 1968. Medieval reception of classical plant names. In *Actes du XIIe Congrès International d'Histoire des Sciences, 21–31 Août 1968, Paris.* Paris.

———. 1979. Identification of the Plants described by Albertus Magnus' *De vegetabilibus* lib. VI. *Res Publica Litterarum* 2:281–318.

———. 1980a. Albertus Magnus and medieval herbalism. In *Albertus Magnus and the sciences: commemorative essays*, edited by J. A. Weisheipl. Toronto: Pontifical Institute of Mediaeval Studies.

———. 1980b. The botany of St. Albert the Great. In *Albertus Magnus, Doctor Universalis*, edited by G. Meyer and A. Zimmerman. Mainz: Matthiàs Grünewald Verlag.

———. 1999. *Pristina medicamenta: ancient and medieval medical botany.* Aldershot: Ashgate.

Stannard, Jerry, Richard Kay, and Katherine E. Stannard. 1999. *Herbs and herbalism in the Middle Ages and Renaissance.* Aldershot ; Brookfield, Vt., USA: Ashgate Variorum.

Sterelny, Kim. 1999. Species as evolutionary mosaics. In *Species, New interdisciplinary essays*, edited by R. A. Wilson. Cambridge, MA: Bradford/MIT Press.

Stevens, Peter F. 1994. *The development of biological systematics: Antoine-Laurent de Jussieu, nature, and the natural system.* New York: Columbia University Press.

Stevens, Peter F. 1997. J. D. Hooker, George Bentham, Asa Gray and Ferdinand Mueller on species limits in theory and practice: A mid-nineteenth century debate and its repercussions. *Historical Records of Australian Science* 11 (3):345–370.

Stresemann, Erwin. 1975. *Ornithology from Aristotle to the present.* Translated by H. J. Epstein and C. Epstein. Cambridge MA: Harvard University Press.

Strickland, Hugh. E., John Phillips, John Richardson, Richard Owen, Leonard Jenyns, William J. Broderip, John S. Henslow, William E. Shuckard, George R. Waterhouse, William Yarrell, Charles R. Darwin, and John O. Westwood. 1843. Report of a committee appointed "to consider of the rules by which the nomenclature of zoology may be established on a uniform and permanent basis". *Report of the British Association for the Advancement of Science for 1842*:105–121.

Temkin, Oswei. 1959. The idea of descent in post-Romantic German biology, 1848–1858. In *Forerunners of Darwin 1745–1859*, edited by B. Glass, O. Temkin and W. L. Straus Jr. Baltimore: Johns Hopkins University Press.

Templeton, Alan R. 1989. The meaning of species and speciation: A genetic perspective. In *Speciation and its consequences*, edited by D. Otte and J. Endler. Sunderland, MA: Sinauer.

Terrall, Mary. 2002. *The man who flattened the earth: Maupertuis and the sciences in the enlightenment.* Chicago: The University of Chicago Press.

Thompson, J. Arthur. 1934. *Biology for everyman.* 2 vols. London: J. M. Dent.

Tournefort, Joseph Pitton, de. 1716–30. *The compleat herbal: or, the botanical institutions of Mr. Tournefort. translated from the original Latin. With large additions from Ray, Gerarde, Parkinson, and others. To which are added, two alphabetical indexes. Illustrated with about five hundred copper plates. With a short account of the life and writings of the author.* Translated by J. Martyn. London: printed for R. Bonwicke, Tim. Goodwin, John Walthoe, S. Wotton, Sam. Manship [and 5 others in London]. Original edition, *Institutiones rei herbariae*, 1700.

Trémaux, Pierre. 1865. *Origin et transformations de l'homme et des autres êtres.* Paris: L. Hachette.

———. 1878. *Origine des Espèces et de l'Homme, avec les causes de Fixité et de Transformation, et Principe Universel du Mouvement et de la Vie ou loi des Transmissions de Force.* 4th ed. Bale, Switzerland; Paris: Librairie Française; Sagnier.

Trewavas, E. 1973. What Tate Regan said in 1925. *Systematic Zoology* 22:92–93.

Turesson, Göte. 1922. The species and variety as ecological units. *Hereditas* 3:10–113.

———. 1929. Zur natur und begrenzung der artenheiten. *Hereditas* 12:323–334.

Van Valen, L. 1976. Ecological species, multispecies, and oaks. *Taxon* 25:233–239.

von Buch, Leopold. 1877. *Gesammelte Schriften.* Edited by J. Ewald, J. Roth and W. Dames. Vol. 3. Berlin: G. Reimer.

Vrana, P., and Ward Wheeler. 1992. Individual organisms as terminal entities: Laying the species problem to rest. *Cladistics* 8:67–72.

Wagner, Warren H. 1983. Reticulistics: The recognition of hybrids and their role in cladistics and classification. In *Advances in cladistics*, edited by N. I. Platnick and V. A. Funk. New York: Columbia Univ. Press.

Wallace, Alfred Russel. 1858. Note on the theory of permanent and geographical varieties. *The zoologist* 16:5887–5888.

———. 1870. *Contributions to the theory of natural selection: a series of essays.* London: Macmillan.

———. 1889. *Darwinism: an exposition of the theory of natural selection, with some of its applications.* London: Macmillan.

Waples, R S. 1991. Pacific salmon, *Oncorhynchus* spp., and the definition of 'species' under the Endangered Species Act. *Marine Fisheries Review* 53:11–22.

Wasmann, Erich. 1910. *Modern Biology and the Theory of Evolution.* Translated by A. M. Buchanan. 3rd ed. London: Kegan Paul, Trench, Trübner. Original edition, 1906.

Weismann, August. 1904. *The evolution theory.* Translated by J. A. Thompson and M. R. Thompson. 2 vols. London: Edward Arnold.

Wells, George Albert. 1978. *Goethe and the development of science, 1750-1900.* Alphen aan den Rijn: Sijthoff & Noordhoff.

Whately, Richard. 1875. *Elements of logic.* Ninth (octavo) ed. London: Longmans, Green & Co. Original edition, 1826.

Wheeler, Q. D. 1999. Why the phylogenetic species concept?—elementary. *Journal of Nematology* 31 (2):134–141.

Wheeler, Quentin D., and Rudolf Meier, eds. 2000. *Species concepts and phylogenetic theory: a debate.* New York: Columbia University Press.

Wheeler, Quentin D., and Norman I. Platnick. 2000. The phylogenetic species concept (*sensu* Wheeler and Platnick). In *Species concepts and phylogenetic theory: A debate*, edited by Q. D. Wheeler and R. Meier. New York: Columbia University Press.

Whewell, William. 1837. *History of the Inductive Sciences.* London: Parker.

———. 1858. *The History of Scientific Ideas.* 2 vols. London: J. W. Parker & Son.

Wiley, E. O. 1978. The evolutionary species concept reconsidered. *Systematic Zoology* 27:17–26.

———. 1981. Remarks on Willis' species concept. *Systematic Zoology* 30:86–87.

Wiley, E. O., and Richard L. Mayden. 2000. The evolutionary species concept. In *Species concepts and phylogenetic theory: A debate*, edited by Q. D. Wheeler and R. Meier. New York: Columbia University Press.

Wilkins, John. 1668. *An essay towards a real character, and a philosophical language. By John Wilkins.* London: printed for Sa: Gellibrand, and for John Martyn printer to the Royal Society.

Wilkins, John S. 2003. How to be a chaste species pluralist-realist: The origins of species modes and the Synapomorphic Species Concept. *Biology and Philosophy* 18:621–638.

———. 2005. A scientific modern amongst medieval species. *University of Queensland Historical Proceedings* 16:1–5.

———. 2006. Species, Kinds, and Evolution. *Reports of the National Center for Science Education* 26 (4):36–45.

Willmann, Rainer. 1985a. *Die Art in Raum und Zeit*. Berlin: Paul Parey Verlag.

———. 1985b. Reproductive isolation and the limits of the species in time. *Cladistics* 2:336–338.

———. 1997. Phylogeny and the consequences of molecular systematics. In *Ephemeroptera and Plecoptera: Biology—Ecology—Systematics*, edited by P. Landolt and M. Satori. Fribourg: MTL.

Wilson, John G. 2000. *The forgotten naturalist : in search of Alfred Russel Wallace*. Melbourne: Australian Scholarly Publishing.

Winsor, Mary Pickard. 1979. Louis Agassiz and the species question. *Studies in History of Biology* 3:89–117.

———. 1991. *Reading the shape of nature: comparative zoology at the Agassiz Museum*. Chicago: University of Chicago Press.

———. 2000. Species, demes, and the Omega Taxonomy: Gilmour and The New Systematics. *Biology and Philosophy* 15 (3):349–388.

———. 2001. Cain on Linnaeus: the scientist–historian as unanalysed entity. *Studies in the History and Philosophy of the Biological and Biomedical Sciences* 32 (2):239–254.

———. 2003. Non-essentialist methods in pre-Darwinian taxonomy. *Biology & Philosophy* 18:387–400.

———. 2004. Setting up milestones: Sneath on Adanson and Mayr on Darwin. In *Milestones in Systematics: Essays from a symposium held within the 3rd Systematics Association Biennial Meeting, September 2001*, edited by D. M. Williams and P. L. Forey. London: Systematics Association.

———. 2006a. Linnaeus' biology was not essentialist. *Annals of the Missouri Botanical Garden* 93 (1):2–7.

———. 2006b. The Creation of the Essentialism Story: An Exercise in Metahistory. *Hist. Phil. Life Sci.* 28:149–174.

Wood, Casey A., and Florence Marjorie Fyfe, eds. 1943. *The art of falconry, being the De arte venandi cum avibus of Frederick II of Hohenstaufen*. Stanford University, CA., London: Stanford University Press; H. Milford.

Wood, Todd Charles. 2007. Bishop John Wilkins, F.R.S. (1614–1672) and his discussion of Noah's Ark *Occasional Papers of the BSG* 9:1–9.

Woodger, J. H. 1937. *The axiomatic method in biology*. Cambridge UK: Cambridge University Press.

Wu, Chung-I. 2001a. Genes and speciation. *Journal of Evolutionary Biology* 14 (6):889–891.

———. 2001b. The genic view of the process of speciation. *Journal of Evolutionary Biology* 14:851-865.

Index

Z